Contents

PREFACE TO THE UNITED STATES EDITION

February, 1977

We are publishing this book in the USA and Canada because it is the best simple guide available on the critical path method of network planning for project management. It is short but complete. The writing is crisp and clear. Over the past five years more than two thousand seminar participants in the United States have found it to be an excellent introduction to network planning. It has been reviewed and praised by experts as well as novices.

Analysis Bar Charting uses the precedence notation. Precedence diagrams (activity on node) are simpler to create and easier to understand than arrow diagrams (activity on arrow) for the following reasons:

1. They are more natural, like flowcharts.
2. They do not require dummy activities.
3. Their numbering system does not use the I-J notation.

Even if the network is going to be entered into a computer, the precedence diagraming approach or *Analysis Bar Charting* is an easy way to start.

Network techniques can be used for planning a wide variety of both large and small projects. In fact, networks are really just an organized way of thinking about complex problems. The sequential logic of network planning is applied common sense and there is no reason to complicate it with fancy terms or computer jargon. The beauty of *Analysis Bar Charting* is in its simplicity.

Networks are plans for the accomplishment of projects which usually involve a group of people. Therefore networks are mediums of communication. How effective they are as plans depends on how effective they are in communicating to the entire organization. A simple approach and an excellent communication device is to create a network using 3x5 cards for activities taped to a board or wall. These networks may be transferred to paper by using pressure sensitive mylar labels and drafting paper. This demonstrates another advantage of the *Analysis Bar Charting* approach over conventional arrow diagraming.

Most of the readers of this book will probably be newcomers to the world of network planning and project management. I wish you the best of success in using your new skill.

Robert Youker
Bethesda, Maryland

PREFACE TO THE SECOND EDITION.

Analysis Bar Charting was first published in 1969 to satisfy the need for a project planning method that was simpler to learn and use than conventional network techniques, such as Critical Path Analysis.

Since then there have been many applications of ABC in a wide variety of projects, from the marketing of food to the overhaul of heavy engineering plant.

One hesitates to claim 'right first time', but this edition contains no alterations to the previous text. Five more years experience reinforces my judgement about the needs of managers in project planning and control.

I suppose I have been privileged, over the years, to introduce network analysis in its critical path analysis form to several thousand managers. As seminars and courses of one sort or another went by, I found that for most of the managers on these courses there were several things wrong with CPA.

First, the symbols used in CPA are the wrong way round; CPA uses an arrow to represent a job and connects jobs together with circles, called events. This is against nature. Most people naturally want to put jobs into 'boxes' and connect them together with arrows.

This opinion is based on a sizeable sample of 'network virgins', managers unsullied by conventional symbolism. And I am not alone – systems which reverse the symbols have grown. Collectively, they are called 'Activity-on-Node' systems.

I will not develop arguments for and against activity-on-node systems. Technically, there are advantages and disadvantages both ways. To me, the clinching factor is that most people coming new to networks put jobs into boxes naturally. I have had to correct this 'error' hundreds of times during training sessions.

Secondly, the thing has become too complicated and has become overloaded with jargon. Surely we can programme work and assign resources, even in complex situations, without using complex ways of expressing ourselves and without introducing a new language. It is this mystique that has caused the failure of many attempts to use networks. Network planning and control has its pay-off on the shop floor, on the construction site or in the brand manager's office. If it is prevented from getting there by complication or jargon and over-elaboration it will not work.

Thirdly, there is the matter of display. If networks are to be used manually for scheduling and resource assignment they need to be converted into bar charts. Many many hours are spent annually on this chore. If we are going to end up with a bar chart, why on earth don't we start with one. The objections to planning complex projects using time-scaled bar charts are well known, but if the methods of CPA are used combined with activity-on-node symbolism, all these objectives can be overcome.

A final point is that the artificial division between 'one-offs' and repetitive work needs to be broken down. Analysis Bar Charting recognises that 'line of balance' for repetitive work situations is only a particular application of a network method. Why use two techniques when one will do?

These are the major factors I have considered in developing Analysis Bar Charting. To summarise:—

ABC uses the 'natural' symbolism of jobs in boxes.
ABC avoids jargon and is simple.
ABC starts with bar charts and avoids conversion.
ABC keeps the logic of the network analysis approach.
ABC deals with repetitive manufacture.

ABC is not designed for the planning and control of large and complex projects for which conventional network methods, backed by computers, are well established. It is a simplified network method primarily intended for manual use.

I have continued to find that managers in marketing, production, construction and administration need a method which keeps the essential values of conventional network techniques but which is simpler to teach and apply and still produces results.

I called this method 'Analysis Bar Charting' because it is based on bar charts, uses an analytical approach, and really is as simple as A.B.C.

1975 J.E.M.

I

Introduction

THE NETWORK FAMILY

The manager who is looking for a technique to help him to plan and control a project is faced with a bewildering family of network methods, marketed under well over a hundred brand names. The exact identity of the father of this family is disputed but there is a high probability that it was PERT (Programme Evaluation and Review Technique). This network method, used on the Polaris programme by the U.S. Navy in 1958, was certainly the first serious attempt to use networks on a large scale. The mother of the American branch of the family was probably CPM, the Critical Path Method of Du Pont. This proved to be a fertile union and the American branch of the network family expanded rapidly. This expansion continues.

By comparison, the British branch got off to a slow start, possibly due to a less healthy management climate; it was founded at about the same time, the nameless father being the Central Electricity Generating Board's network technique, used to plan and control the overhaul of capital equipment.

All these techniques show a strong family likeness. The differences that exist are mainly differences of detail and emphasis. The techniques have survived and prospered because they all bring the same powerful discipline to bear on the task of planning and controlling projects. Their power lies in this basic discipline and not in any of the elaborate embellishments some of them exhibit.

AN APPLICATION OF COMMON SENSE

Network techniques are not new; let no one pretend that here is a fundamentally different approach to project management which, at one stroke, will abolish all the confusion and mistakes of the past. Most managers have been using a form of network analysis for years, although they have not called it by that name. They have probably called it common sense. Before the advent of work study, managers called that common sense, too, and some still do.

COMPLEXITY AND UNCERTAINTY

Common sense, however, is not enough, even assuming there is plenty of it available. Managers today face a great increase in the complexity of their work and, because they are usually dealing with the future, they also face uncertainty. Network techniques were designed specifically to deal with these two factors. They provide the assistance a manager needs when he must go beyond the limit of his mental planning capacity, when he is faced with the task of defining the complex relationships that exist in sequence and time between the many jobs involved in any project he is planning.

When a project is under way and the inevitable deviations from plans occur, network techniques help the manager to determine the importance of these deviations and to take effective corrective action.

COMPLICATION

As network techniques multiplied and developed, their complexity increased. Interesting new methods of manipulation were devised and a whole dictionary of terms came into being. So great was the fascination of the technique that in some cases its purpose, that of getting things done in the right order at the right time, was forgotten. Any elaboration or sophistication which does not contribute to this purpose is, at best, useless.

This preoccupation with technique has led to more than one failure in application. Concern with the more complex manipulations, and admiration of the intricate beauty of some of the elegant calculations devised by mathematical minds, has caused some managers to forget that the strength of networks lies in the logical approach to planning and control that they compel.

Networks are not for decoration, they are for use. As plans they must be crisp, taut representations of the work that has to be carried out. As a means of control they must be flexible, fast-reacting analytical tools. They can be neither if they are full of unnecessary details and complications.

JARGON

The vocabulary of network methods has also grown at an alarming rate. It has apparently been necessary to invent jargon terms for things that already have quite adequate descriptions. Why tell a foreman that the job he is doing has 'negative float' when we mean that he should have finished it yesterday?

It is no answer to say that the use of jargon can be confined to the people who understand it; for example, that a critical path analyst will talk about negative float to his colleagues but not to the foreman. Planners and 'doers' should speak the same language. There are enough differences between them without adding to the difficulties.

USE BY MANAGEMENT

Network analysis is classed as an operational research technique by some authors. This classification might have been valid at one time but it is no longer; the research finished some time ago. Today, network analysis is not a technique to be left in the hands of a few high priests. It is firmly in the hands of managers at all levels in an organisation. The high priests, however, are reluctant to relinquish their hold and some have tried to hide the simplicity of what they are doing under a cloak of mystique.

This is not to say that the research that has been carried out on the more advanced aspects of network theory has been a waste of time. There are applications where the most sophisticated methods are valuable. With the increasing use of computers to process network data, some degree of specialisation becomes necessary. The point to be made is that networks should be kept as simple as possible, both in their form and content. Analysis Bar Charting (ABC) is a network method which has been designed to avoid unnecessary complication and jargon. It is for use in the planning and control of projects where a computer is not used. Its design has been based on the need for understanding and acceptance on the shop floor, on the construction site and in the sales office because, in the end, these are where the work is done. It will help to get it done in the right order and at the right time.

THE BASIC STEPS

Apart from Chapter 8, which deals with the use of ABC for the planning and control of certain types of repetitive work, this book is concerned with the planning and control of projects. The word 'project' can be interpreted widely; it means any task that has a beginning and an end, and during which each job occurs once only.

In these conditions management has two functions to perform. It has to plan and control—to plan the work and work the plan.

PLANNING THE WORK

The objective of planning is to produce a timetable of work, with each job allocated a start date and a finish date and with the assurance that the things necessary to do each job will be available when required.

ABC accomplishes this in a number of steps which are described in detail in the chapters which follow. These steps are: *logic, timings, analysis*, and *scheduling*.

LOGIC

In the first step the individual jobs which make up the project are arranged in their correct order. During this step no consideration is given to how long the jobs will take, neither are the resources necessary to carry out the job considered. Both these steps come later.

The method used to get the jobs into their right order is graphicai and involves the construction of a network. At the end of this step management will have on paper the first representation of its plan. The jobs will be arranged in their correct sequence, and how they depend on one another will be shown.

TIMINGS

The second step is the estimation of the duration of each job and the placing of this information against the representation of each job on the network. Resources necessary to carry out the jobs are considered only to a limited extent at this stage.

ANALYSIS

The network is now analysed to find out two things. In all projects there is one sequence of jobs that is longer than any other and this determines how long the project will take. The first task in the analysis stage is to find out what this sequence is. It will be called the 'critical path through the network' and the jobs in it will be called 'critical jobs'. The analysis will produce start and finish dates for these critical jobs.

All the other jobs are not critical; they will have time to spare. The analysis will determine the dates between which these non-critical jobs can be performed.

SCHEDULING

This is the stage at which the resources are considered in detail. Scheduling involves the taking of decisions about the starting and finishing dates of the non-critical jobs in order to produce a timetable for the project.

WORKING THE PLAN

The second function that management has to perform is to ensure that the work takes place according to the plan. It would be unrealistic to ignore the fact that it seldom does. If deviations from the original plan are so large that they cannot be corrected, this control phase may involve replanning to take account of the changed circumstances.

SUMMARY

The main value of network analysis is the logic and discipline it brings to planning and control. Therefore, it should be kept as simple as possible. It is not a specialist technique but should be understood and used by management at all levels.

ABC is a simple and practical network technique for use when a computer is not necessary. It is applied in both the planning and control phases of a project. In the planning phase, four basic steps —logic, timings, analysis, and scheduling—are taken.

2

Logic

PURPOSE OF THE LOGIC STAGE

This stage is concerned *only* with finding out the order in which jobs *must* be done. Dismiss for the moment any consideration of how long the jobs are going to take and what resources will be required for them.

SYMBOLS USED

JOBS

Bars are used to represent jobs, but because we are not yet concerned with time they can all be the same length and will look like

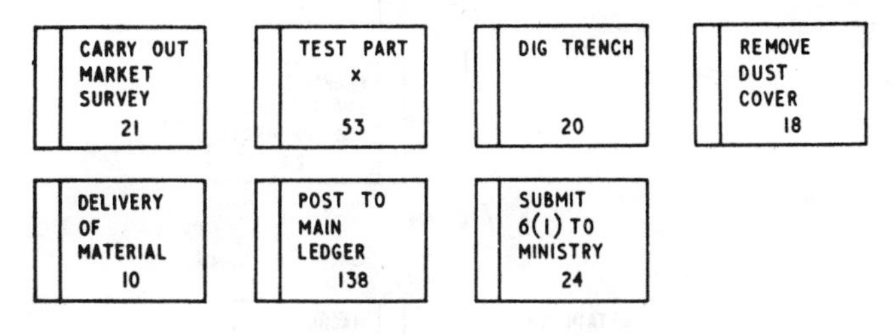

Fig. 2.1

boxes. The description of each job is placed in its box and it can be identified also by giving it a job number, as shown in Fig. 2.1.

The word 'job' also applies to waiting for other people's jobs to take place and includes, for example, delivery periods from suppliers. Waiting for processes such as paint drying or concrete curing

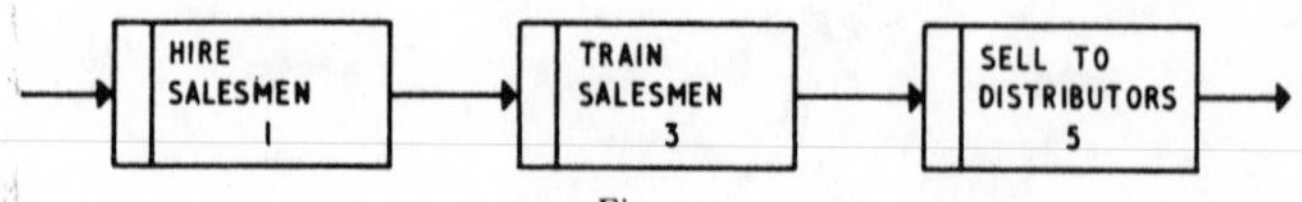

Fig. 2.2

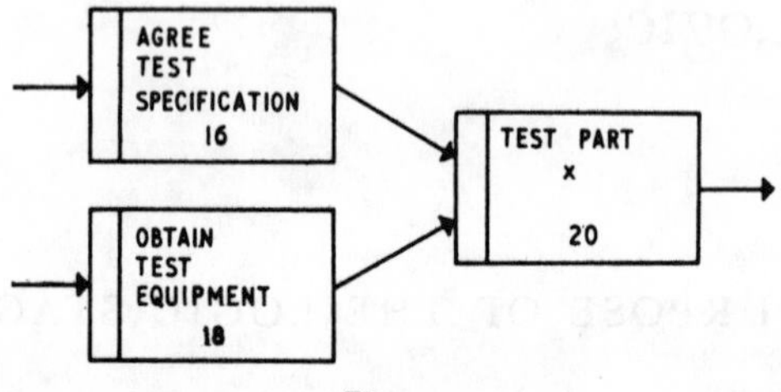

Fig. 2.3

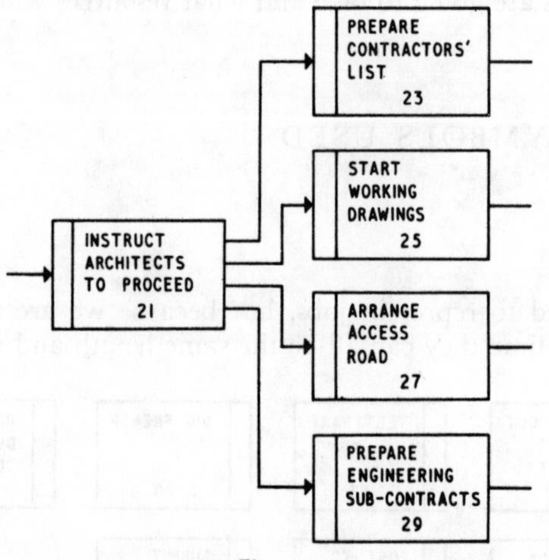

Fig. 2.4

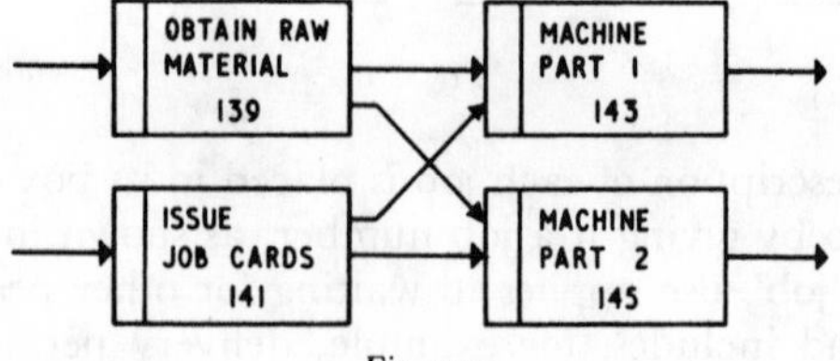

Fig. 2.5

are also classed as jobs. Anything that is necessary to the project and takes up time counts as a job. The compartment on the left of the box will be used for showing how long the job will take, as explained in Chapter 3.

JOB CONNECTIONS

Arrows are used to show how jobs are related. The length of the arrows means nothing, so they can be drawn in any way that is convenient. Figure 2.2, for example, illustrates a straight sequence—as soon as one job has finished the next can begin. Training salesmen depends on hiring them, selling to distributors depends on salesmen being fully trained.

In Fig. 2.3 two jobs must be completed before a third can begin. Before part x can be tested the test specification must be agreed and the test equipment obtained. In Fig. 2.4 only after the architects have been instructed to proceed can the four subsequent jobs be started. Note that Fig. 2.4 does not mean that the four jobs *will* be started immediately after instructing the architects has finished. It means that they *can* be started then.

In some cases the arrows will cross (Fig. 2.5) but this is not important as long as it is clear where each arrow leads. A logic network for the whole project is built up in this way using only these two symbols. Note that we use the term 'network' to describe the representation of a project using boxes and arrows.

NUMBERING OF JOBS

The numbers of jobs only serve to identify them. It is convenient to number in an ascending sequence and also to leave gaps in the numbering so that amendments to the network can be made without disturbing the sequence.

'START' AND 'FINISH' BOXES

It is necessary to begin the logic network with a single 'start' box and to end it with a box labelled 'finish'. The use of these boxes will be described in Chapter 4, which deals with analysis of the network.

A simple logic network, which will be called the 'standard'

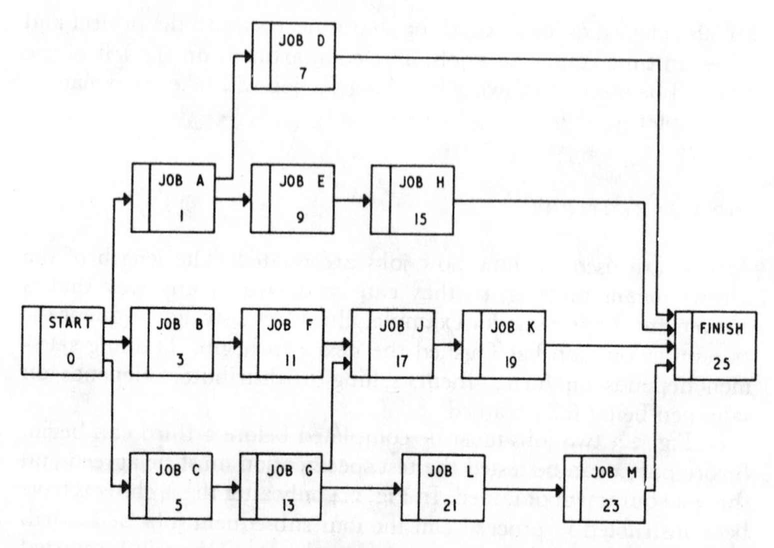

Fig. 2.6. The 'standard' network

network, is shown in Fig. 2.6. This is the network which will be used in subsequent chapters to illustrate the remaining steps of the planning phase.

CONSTRUCTION OF NETWORKS

There are two ways of constructing the logic network. In the first, all the jobs that go to make up the project are listed before the network is started. Jobs can be listed and allocated job numbers by department or by area of work, as shown in Tables 2.1 and 2.2.

Table 2.1. TEST MARKET PROJECT

Market Research Dept.	*Production Dept.*	*Marketing Dept.*	*Management*
M1 Detailed arrangements with agency M2 Draw up contract M3 etc.	P1 Install machine P2 Trial run P3 etc.	Mk1 Arrange test market Mk 2 Agree sampling procedure	Mt1 Authorise test market Mt2 Approve budgets

Table 2.2. CONSTRUCTION PROJECT

Architects	A1 A2 A3	Site investigation Start preliminary design sketches etc.
Quantity Surveyors	Q1 Q2 Q3	Preliminary bills of quantities Confirm estimates etc.
Engineers	E1 E2 E3	Determine services required First estimate of costs etc.

In the second method the jobs are not listed but the network is used as a means of 'thinking aloud'. The network can be drawn commencing at the start box and developed from the beginning of the project. An alternative method sometimes used is to start drawing the network from the finish box and to work backwards. If this procedure is followed, one is forced to ask the question: 'What must we have done *before* we can start this job?' This is a stringent test of logic and is the main advantage of starting at the finish. In a manufacturing example, the reasoning might be as in Fig. 2.7.

In this method the job numbers are allocated after the network is drawn. The method chosen is a matter of personal preference. The author, for example, normally likes to start at the beginning of

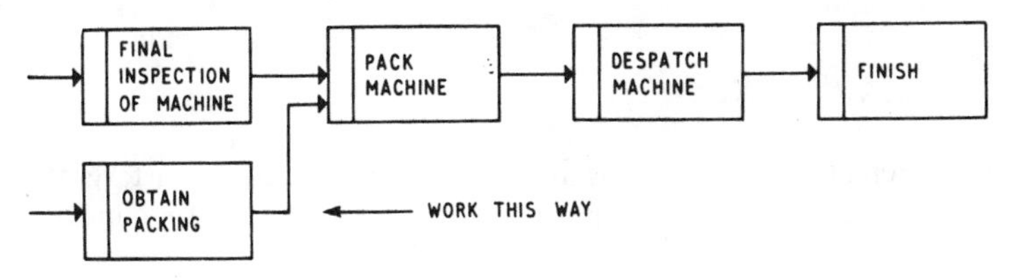

Fig. 2.7

a project, but has been forced to start at the finish in projects when the logic has been particularly complex. Again, some people always compile lists of jobs, while others jump in at the deep end and start drawing without this initial step. Lists of jobs are probably advisable if the project is complex and the inter-relationships of the jobs not easy to see.

If pieces of card are cut to the size of the job boxes, say $1\frac{1}{2}$ in. by 1 in., and the job descriptions are written on them, the network can be constructed by arranging the cards on a desk top. This jigsaw puzzle method is preferred by some managers as it saves much of the drawing effort in the initial stages. When the cards have been arranged to everyone's satisfaction the network is then drawn.

ERRORS IN LOGIC

LOOPING

If a loop is drawn in the network this is clearly a logical error, because once into the loop one can never get out and the project will go on for ever. This may seem to be an accurate representation

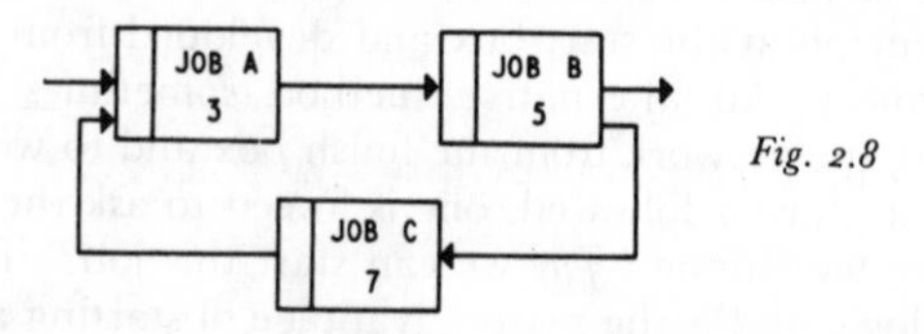

Fig. 2.8

in some cases, but a project should not be *planned* that way! In the example in Fig. 2.8 the loop is easy to see. In a larger network with many jobs and arrows a loop may pass unnoticed.

DANGLING

A job which is left dangling in mid-air has usually been forgotten. In Fig. 2.9, Job 25, if it is part of the project, should be connected

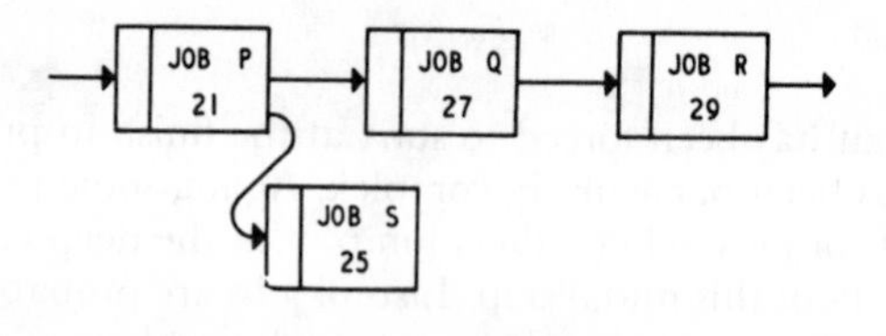

Fig. 2.9

back into the network somewhere, even if it is only to the finish box. Unless this is done there will be two 'ends' to the network, which will cause trouble in the analysis stage.

OVERLAPPING JOBS

Up to now we have considered only those jobs that take place in neat boxes; when one finishes the next starts. Many overlaps occur in practice and a way of representing this situation is required.

Consider the case of production followed by distribution. If we use the convention so far adopted we would have a diagram such as

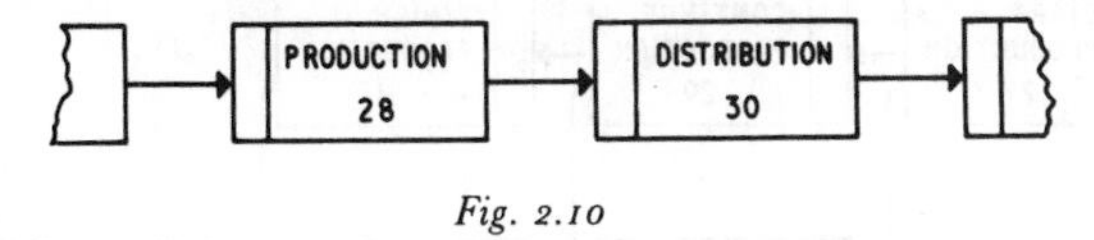

Fig. 2.10

Fig. 2.10. However, it might not be necessary to wait until production has finished before beginning distribution. The convention used is shown in Fig. 2.11. In practice, the arrows marked T_1 and T_2

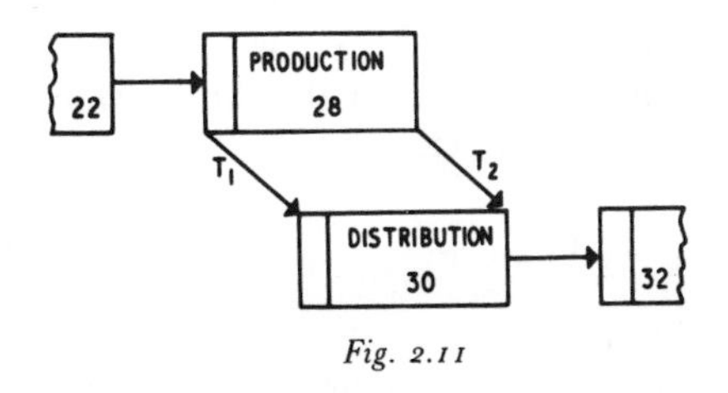

Fig. 2.11

would have times placed against them. T_1 is the time necessary between *start* of production and *start* of distribution; T_2 is the time

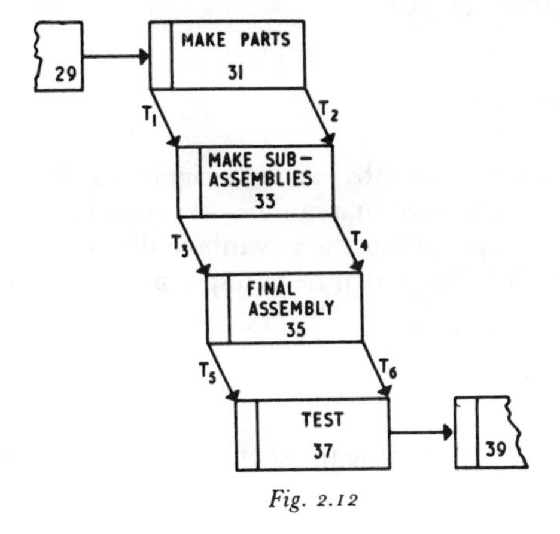

Fig. 2.12

necessary between the *end* of production and the *end* of distribution.

If more than two jobs overlap, the diagram is simply extended as shown in Fig. 2.12. This method of representing overlapping jobs will be dealt with in more detail in Chapters 3 and 4.

An alternative way of representing overlapping jobs is to split them up into elements. Consider the production and distribution

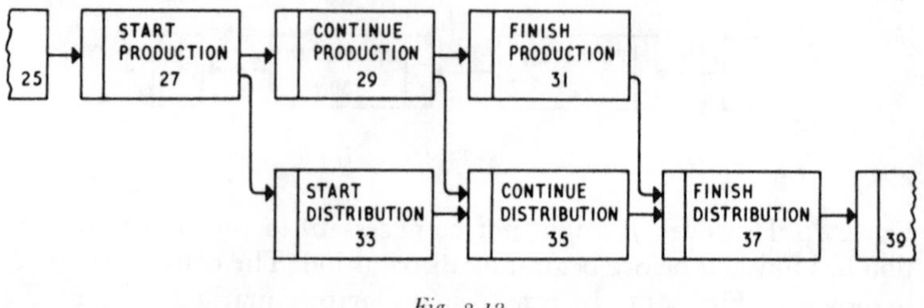

Fig. 2.13

overlap situation; this could be drawn as shown in Fig. 2.13. This method involves estimating times for the three elements of each job and will also be dealt with later.

LAYOUT

This is again a matter of personal taste. Any layout is acceptable as long as it is neat and can be read easily. Several categories of layout have emerged over the years.

ZONED NETWORKS

Where responsibilities for jobs can be assigned to departments the network can be constructed in a number of zones. Figure 2.14 shows a network of this type. It has the advantage that everyone can see, very clearly, the jobs for which he is responsible.

CRANKSHAFT

This layout is so called because the shape of the network resembles a crankshaft (Fig. 2.15).

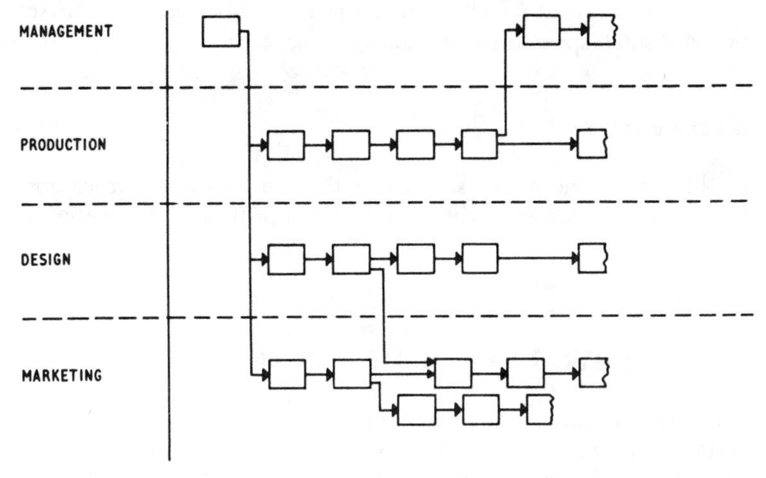

Fig. 2.14. A zoned network

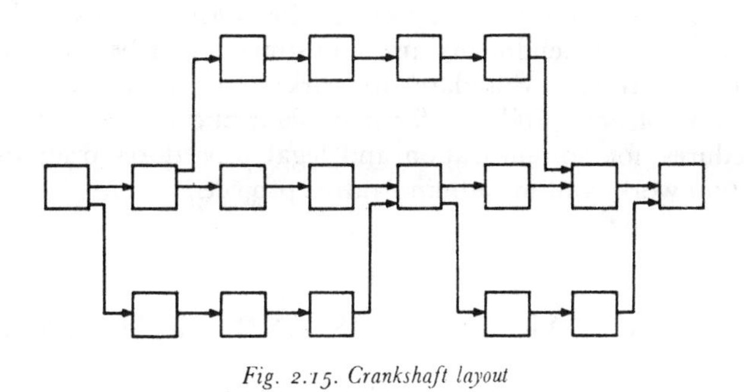

Fig. 2.15. Crankshaft layout

ZIGZAG

A method of layout which enables a large number of jobs to be set out in a small space is called 'zigzag' (Fig. 2.16).

WATERFALL

In this layout the network starts at the top left-hand corner and finishes at the bottom right, giving the appearance of a waterfall (Fig. 2.17).

IMPORTANCE OF LOGIC

Too much attention should not be paid to the layout during the initial construction of the network, which is an exercise in logical thinking, not draughtsmanship. The first attempt will probably look like grandma's knitting, but once the logic is right, layout can be considered and, finally, the network drawn neatly.

STANDARD NETWORKS

When an organisation repeats a project from time to time, standard networks can be used. If the form of the work to be carried out is similar, minor amendments and adjustments can be made to suit particular projects. Standard networks have been used for the launching of new products, for manufacturing projects, for design procedures, for administration and legal procedures prior to construction work, and for administrative projects.

PREPARED GRIDS AND TEMPLATES

To save the job of drawing the boxes, grids can be used with the boxes printed on. All that is then required is to name and number the jobs and draw the connecting arrows. Templates made of transparent material with the boxes cut in them are also useful.

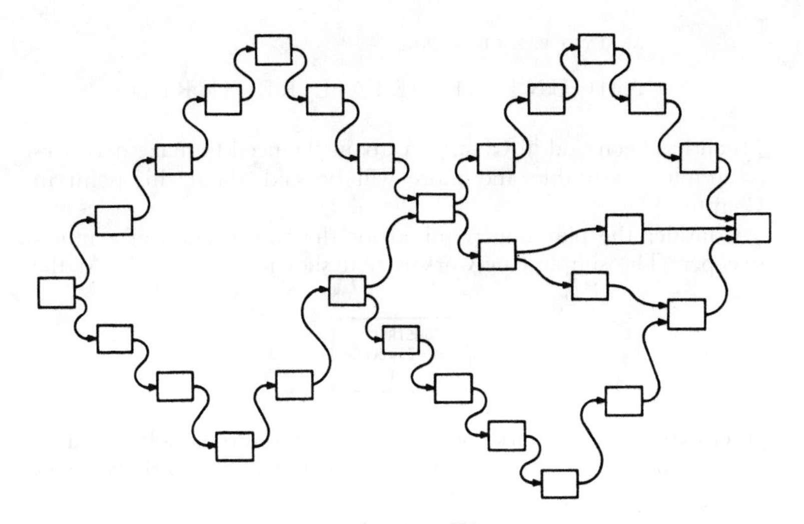

Fig. 2.16. Zigzag layout

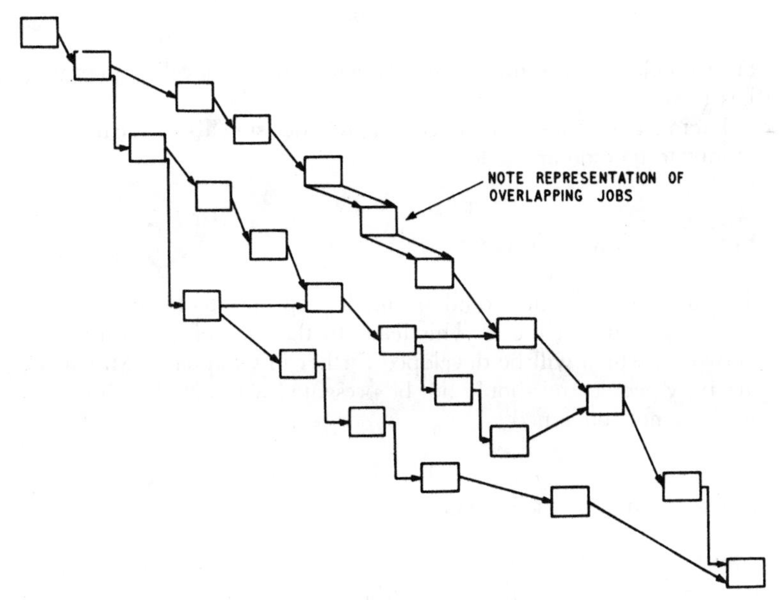

Fig. 2.17. Waterfall layout

AMOUNT OF DETAIL REQUIRED

Much has been said in Chapter 1 about the need to make networks as simple as possible and more will be said about this point in Chapter 9.

Consider the planning required for the construction of a mine-sweeper. The simplest network is that shown in Fig. 2.18. At the

Fig. 2.18

other extreme, a network consisting of many thousand jobs could be drawn, containing jobs such as in Fig. 2.19. Clearly, both extremes

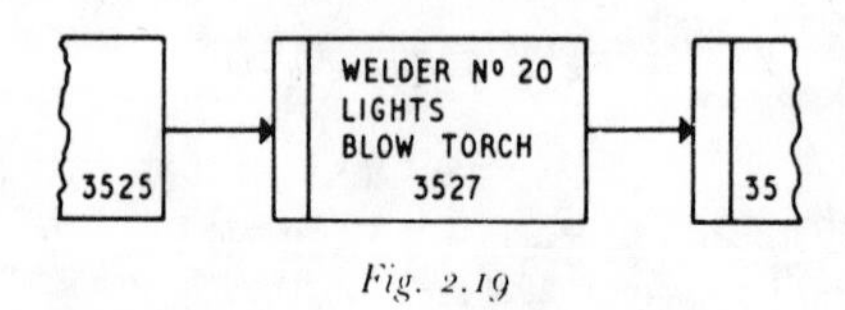

Fig. 2.19

are ridiculous, but somewhere between them the 'level' of planning has to be set.

Factors to be considered in coming to a decision about the number of jobs to include are as follows.

LEVEL OF MANAGEMENT

The degree of detail should be made appropriate to the level of manager who will use it. This leads to the idea of a hierarchy of networks, which will be developed further in Chapter 9. Managers are busy people and should not be presented with detail with which they are not concerned.

DEGREE OF CONTROL REQUIRED

This factor involves consideration of the control procedure that will be adopted, which will be described in Chapter 7. If corrective action is to be taken weekly this gives a good indication of the minimum length of individual jobs to be considered. Apart from a few exceptions, it will probably not be necessary to split the work down to jobs of less than one week's duration.

RESPONSIBILITY

Whenever possible each job should be the responsibility of one person. When responsibility changes, a separate job should be drawn. For example, if the preparation of a document involves both architects and planners, their jobs should be shown separately.

MAXIMUM NUMBER OF JOBS

It has been the author's experience that good control can be obtained with networks limited to about 250 jobs. Some have required considerably less. Above this limit the amount of data that has to be handled becomes too great for manipulation without computer assistance, particularly if things are changing rapidly.

PLANNING CYCLES

Sometimes it is necessary to go through a planning cycle to arrive at a satisfactory network. The pattern is as follows:

Step 1. Draw simple network of, say, 50 jobs making assumptions about interdependence.

Step 2. Expand simple network into greater detail, say 200 jobs. This will test the assumptions made and 'throw up' interdependencies that could not be determined initially.

Step 3. Condense the detailed network to a smaller size respecting the new relationships revealed by Step 2.

WHO DRAWS THE NETWORK?

This will be examined in detail in Chapter 9. At this stage all that will be said is that network construction must be based on a thorough understanding of the task to be carried out and the way the organisation works.

SUMMARY

Logic is the first stage in planning using ABC. The symbols used have been introduced. The method of representing overlapping jobs has been explained. Job lists and numbers can be prepared before the network is drawn. Alternatively, the network can be drawn without these lists, and numbers allocated afterwards. Networks can be drawn from either end. Loops or dangles are not permitted. Layout is a matter of preference; some examples have been given. The amount of detail to be included is important; some guide lines have been explained. Whoever draws the network must understand the task and the organisation.

3
Timings

The second stage in the planning phase is to estimate the duration of each of the jobs. The estimation of job times is a task that must be undertaken whatever the form of planning being used, it is not something that is required only when using network methods.

UNCERTAINTY

Because we are now dealing with the future, uncertainty is bound to exist. In some cases the degree of uncertainty will be great. Who can say what delays will be caused by weather conditions on a construction project? Who can predict what will be found when the pump is dismantled during an overhaul? Who can say what difficulties will arise during the design of the servo-system on the missile? What reliance can be placed on the sub-contractors' delivery promises?

It is management's responsibility to face these uncertainties squarely and to do the best they can about them. Any form of planning involves making decisions based on information that is uncertain. As far as job timings are concerned, all we can do is to use the best estimate available at the time the decision has to be made. Subsequently, more reliable information may become available and the revised timings can be used to replan the parts of the project affected.

REALISTIC TIMINGS NECESSARY

All time estimates should be realistic. The time used should be the time that is most likely to be needed, not an optimistic or a pessimistic time. The difficulty is in determining this most likely time. In most organisations there are a number of sources from which time estimates can be obtained.

WORK MEASUREMENT

Some jobs are easy to estimate because they have been done a number of times in the past and have been measured by one of the work measurement methods. These tend to be production jobs where, in some cases, 'standard times' may be available.

For jobs that have been studied in detail, a performance time may be estimated by the use of synthetics. Where a work study department exists this can often produce realistic time estimates for jobs.

EXPERIENCE

Where such quantified data does not exist, it is necessary to use management experience in the past as a guide to the future. 'Experience' is usually available in two forms; records and memories.

If records of similar work in the past are available, these can often be used as a basis for time estimates. One of the advantages of using network methods is that a record of times achieved is built up, which can then be used to improve the reliability of future estimates. Reasonably reliable estimates can sometimes be obtained by an examination of labour and staff costs on past projects.

Memories are usually not as reliable as records but in the absence of records it is necessary to ask people to try to remember how long a job took 'last time we did it'. It is important to find the man or men who know most about a particular job and to ask them. This raises another important point.

RESPONSIBILITY

It is fatal to impose times on the men responsible for a job. In all cases the job duration should be discussed and agreed with the people who will be carrying it out. Failure to do this will certainly generate resistance. In practice, this often means seeking advice from people quite junior in the management structure, but it is foolish to ignore the vast amount of knowledge that exists at these junior levels. In short, find out who knows most about the technicalities of the job and ask them.

NEW JOBS

Jobs that are entirely new present more difficulty. One method of improving on pure guesses is illustrated by the example that follows.

Suppose that there is a design job in a network (Fig. 3.1) and the designer responsible is loath to commit himself. His task is often

Fig. 3.1

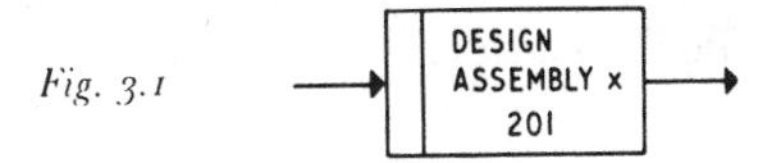

made easier if the job is broken down into its elements and a small network constructed. Suppose that, on examination, the work he has to do turns out to be as shown in Fig. 3.2. Times can then be estimated for each of the jobs and the small network analysed, by the

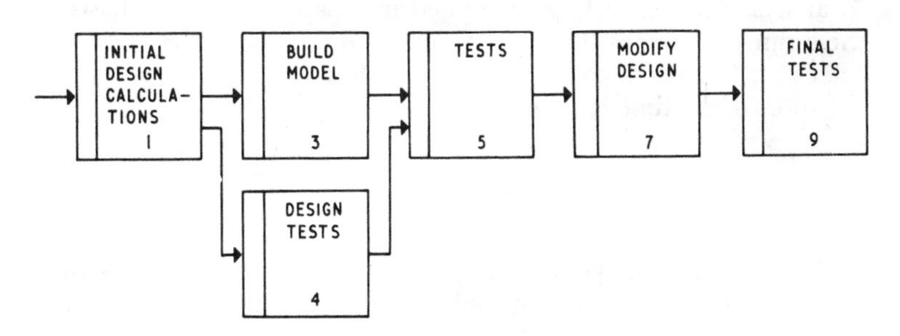

Fig. 3.2

methods to be described in Chapter 4, to arrive at a total time. The reason that this procedure helps is that people find it easier to estimate small jobs than large ones and, secondly, setting out the elements of a job in network form *forces* them to think logically about the work they have to do. In practice, the 'sub-networks' tend to be larger than the one in Fig. 3.2, but about 20 jobs are normally enough.

MULTIPLE TIME ESTIMATES

Some network systems use three time estimates and combine these to arrive at an estimated time that is 'most likely' on a statistical basis. The author does not consider that this is normally worth while but the approach is useful.

When faced with the task of estimating duration time for a job that is difficult, first estimate the probable limits. For example:

Time for Design of Assembly X
Optimistic time = 6 weeks
Pessimistic time = 12 weeks

The optimistic time is the time that would be taken if everything went smoothly. The pessimistic time is the time that would be taken if all the foreseeable difficulties arose. The most likely time will lie somewhere between these two limits and the task of deciding its value is made easier if these limits are established.

REVISION OF JOB TIMES

It is sometimes necessary to make a first estimation of job times and to analyse the network, as described in Chapter 4, on the basis of these times. The jobs that are 'critical' will then be determined and these can be examined more closely with a view to improving the accuracy of the time estimates.

RESOURCES

It is necessary to make some assumptions about the resources that will be used to carry out a job when estimating how long it will take.

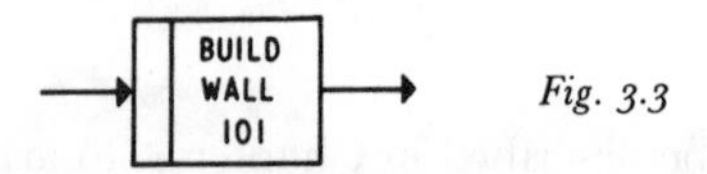

Fig. 3.3

Suppose one of the jobs in a network is that shown in Fig. 3.3. If the work content of building this wall is six man-days, the duration of the job will depend upon the number of bricklayers employed. For one, it will be six days; for two, it will be three days, and so on.

The rule here is that the 'normal' level of resource must be assumed. This assumption will be examined in more detail during the scheduling step, as will be explained in Chapter 6.

HOW TIMES ARE USED

The job times are now placed on the network against the job to

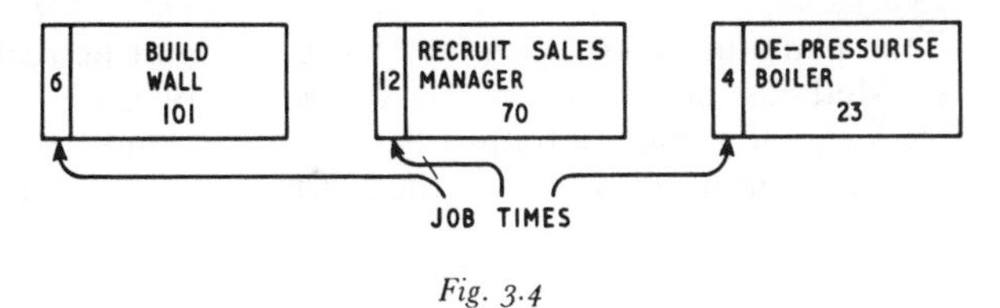

Fig. 3.4

which they refer. The duration box at the left-hand side of the job box is used for this purpose as shown in the examples in Fig. 3.4.

TIME UNITS

These can be working days, weeks, hours, 'shifts' or any other appropriate time units. It is better not to use fractional time units if

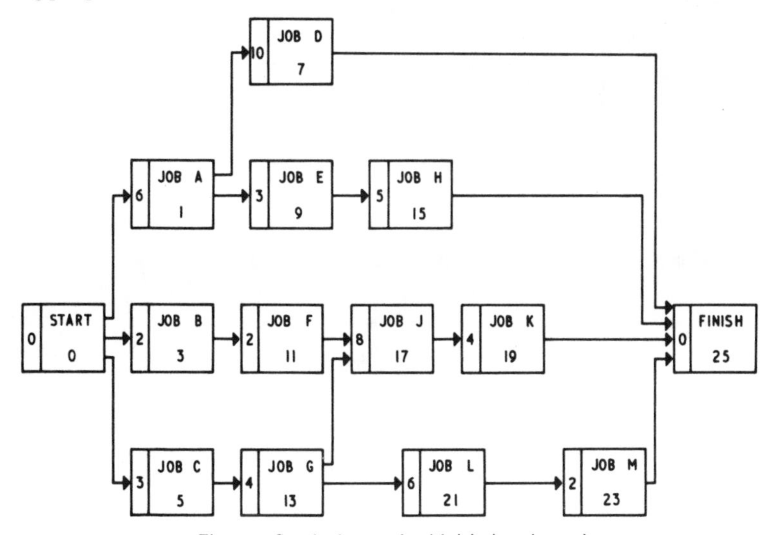

Fig. 3.5. Standard network with job times inserted

they can be avoided, since they make the job of analysis more difficult.

The standard network in Fig. 2.6 becomes as shown in Fig. 3.5 when the job times are added.

SUMMARY

Step 2 of the planning phase is timing. Job times must be realistic estimates. Methods of estimating have been explained. Those responsible for performing a job must be consulted. Time estimates in an appropriate unit are placed in the left-hand compartment of each job box.

4

Analysing the Network

OBJECTIVES OF THE ANALYSIS STAGE

The logic stage put the jobs into their right order and the timing stage gave the durations of each job. The purpose of analysing the network is to determine *when* the individual jobs *can* be run.

In any project there is one sequence of jobs that fixes the duration of the project. Naturally enough, it is the longest sequence and the jobs in it are called 'critical'.

The analysis determines which these jobs are. All the other jobs will have time to spare and the analysis determines exactly how much. In other words, analysis tells us, first, which jobs are critical and when these *must* take place and, secondly, gives the limits for the non-critical jobs, saying when they *can* take place.

HOW START AND FINISH TIMES ARE SHOWN ON THE NETWORK

It will be recalled that so far we have not drawn the boxes to scale. Each job is represented as that in Fig. 4.1. In this figure, Job F,

Fig. 4.1

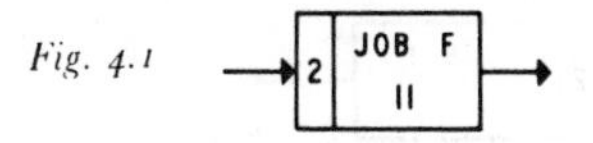

Job number 11, takes 2 time units and its relationship with the other jobs is shown by its position in the network.

Analysis will answer the following questions: How soon can the job start? How late can the job start? How soon can the job finish? How late can the job finish? The answers to these questions will be given on the job box as shown in Fig. 4.2. How these times are determined will now be illustrated using the 'standard' network of Fig. 3.5. For the sake of clarity, job numbers will be omitted.

THE FORWARD PASS

The first step is to make a 'forward pass' through the network from start to finish, adding up the durations to give the *earliest* starts and

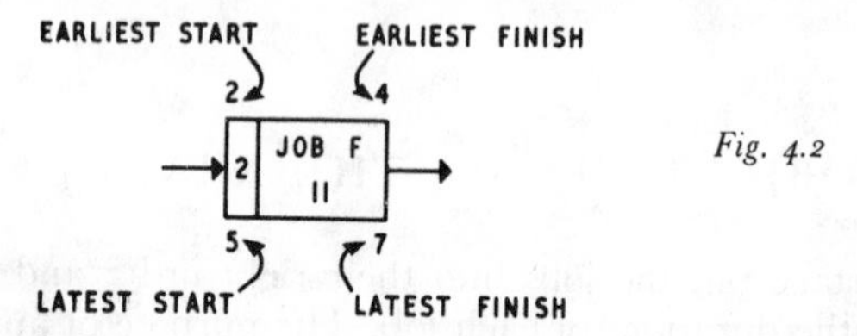

Fig. 4.2

finishes, which are then recorded on the top of the job box as shown in Fig. 4.2.

A further example is shown in Fig. 4.3, where the start box has a duration of zero and is put into the network to provide a common

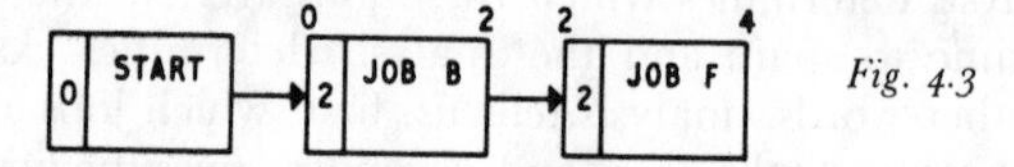

Fig. 4.3

starting point for the jobs that follow. Assume that the time is days. Job B can start at time 0, the beginning of the first day and, since its duration is 2, the earliest it can finish is 2, the end of the second day. From now on we refer to the *end* of days in all cases.

The logic says that Job F can start only when Job B has finished, so its earliest start date is 2. Since it takes 2 days, its earliest finish

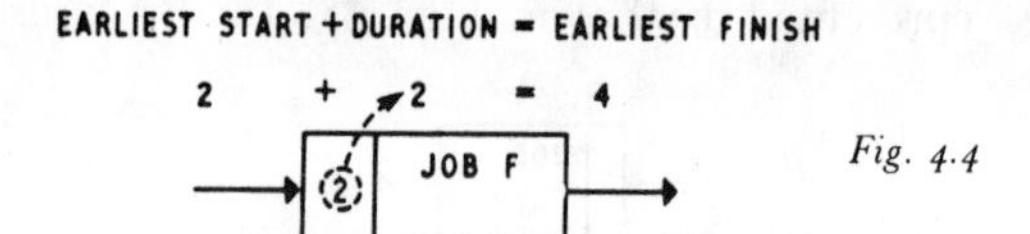

Fig. 4.4

date is 4, as shown in Fig. 4.4. In the same way, the job sequence C–G–J can be analysed as shown in Fig. 4.5.

The logic states that Job J 'depends on' Jobs F and G. The earliest F can finish is 4. The earliest G can finish is 7. Job J, therefore, cannot start until 7 at the earliest, as shown in Fig. 4.6.

It is important to remember that, whenever more than one job

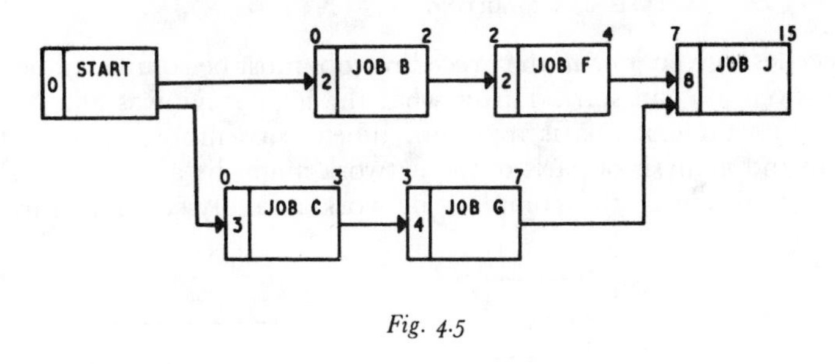

Fig. 4.5

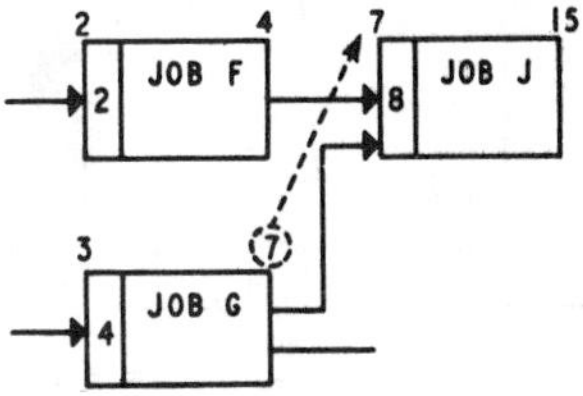

Fig. 4.6. Job J depends on Jobs F and G and so cannot start until they are both finished

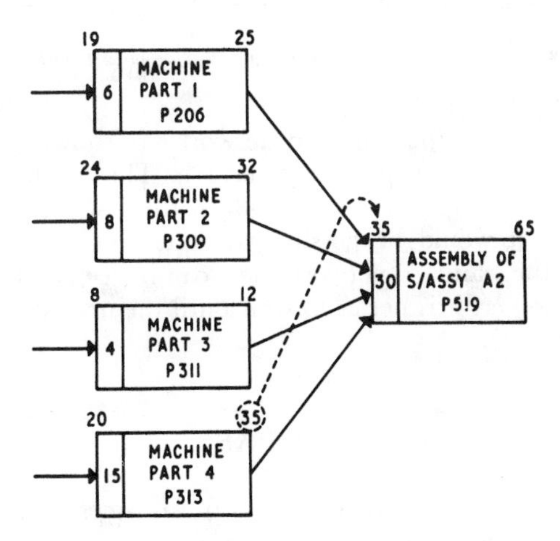

Fig. 4.7. Part of a network showing that assembly cannot begin until all four parts are machined. The dotted line shows how the earliest start for the assembly job is determined

precedes a given job, *all* the preceding jobs must be completed before the given job can start. This is what the logic stage was all about.

For example, if 4 parts are required before assembly can begin, the logic and analysis of parts of the network might be as shown in Fig. 4.7. In this way the complete network is analysed. The earliest

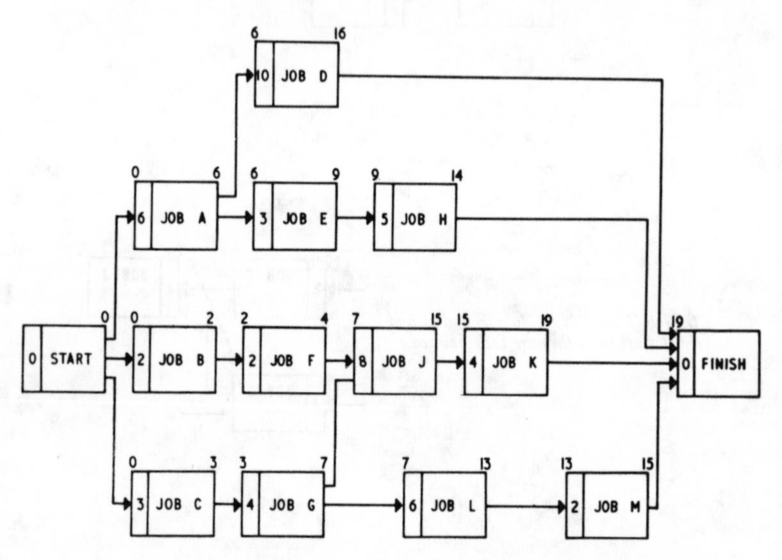

Fig. 4.8. Standard network—result of forward pass showing earliest starts and finishes

starts and finishes for the standard network are shown in Fig. 4.8. The earliest finish to the project is day 19. The finish box has a duration of 0 and simply serves to 'collect' the last jobs together.

At this stage it would be possible to determine which jobs are critical and in the standard network these can be seen easily enough. In general, however, it is better to wait until the next step has been taken.

THE BACKWARD PASS

This step determines the *latest* start and finish times. Starting from the end of the network, duration times are subtracted and latest starts and finishes recorded on the *bottom* of the job box.

The first question that arises is what figure to use for the latest finish to the complete project. The forward pass determined that the project *could* be completed in 19 days.

As a first step this is the figure to use. We may end up having to complete the project in under 19 days, or we may have more than 19 days available. Both of these situations will be examined separately. For the moment assume that we wish to complete the project in 19 days. Figure 4.9 shows how some of the jobs in the standard

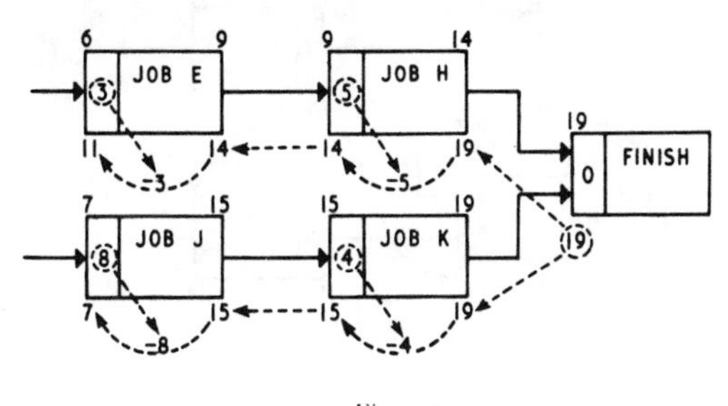

Fig. 4.9

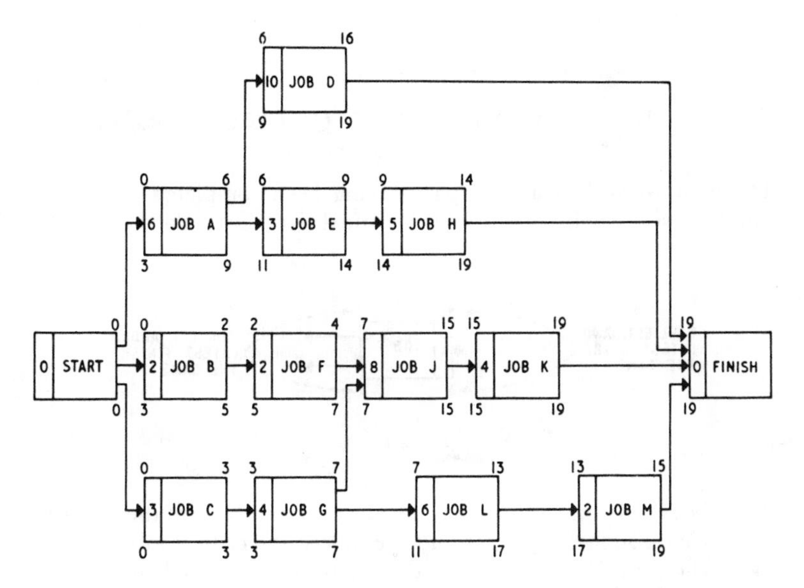

Fig. 4.10. Standard network—result of backward and forward pass showing earliest and latest starts and finishes

network are analysed. The dotted arrows indicate how the times are transferred. The complete project is analysed in Fig. 4.10.

The only difficulty likely to arise here occurs at the junctions. Consider the three jobs A, D and E. The backward pass gives day 9 as the latest start for Job D and day 11 as the latest start for Job E.

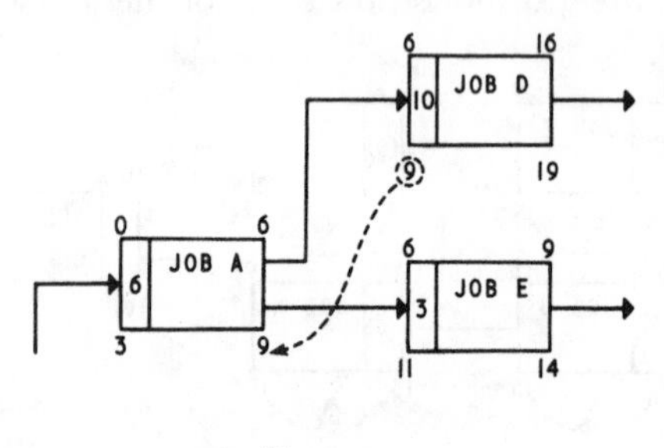

Fig. 4.11

Both D and E depend on A and so the latest that A must be finished is in time for D to begin on day 9. This is shown in Fig. 4.11.

DETERMINING THE CRITICAL PATH

The forward and backward pass having been completed it is now simple to see which of the jobs are critical. In the standard network

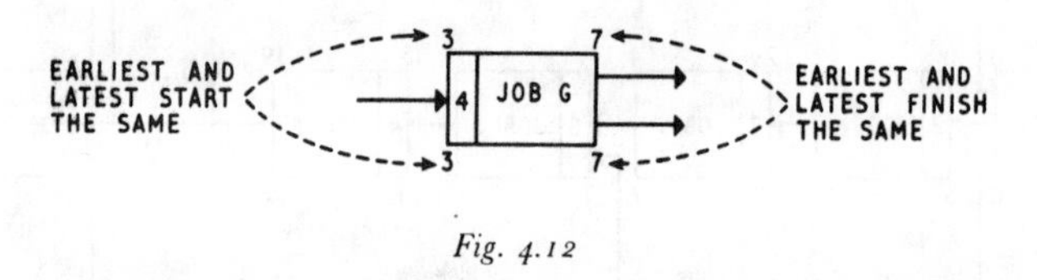

Fig. 4.12

it can be seen by inspection that jobs C, G, J and K form the longest sequence through the project and are indeed the critical jobs. When the earliest and latest starts for these jobs are inspected, it will be seen that they are identical, as are the earliest and latest finishes. For example, in Fig. 4.12, Job G has an earliest *and* latest start of day 3 and an earliest *and* latest finish of day 7. This means it has no time to spare and *must* be run between the end of day 3 and the end of day 7, if the project is to be completed in 19 days.

The critical path can be indicated by the use of colour, usually red, on the arrows; however, in Fig. 4.13 it is shown by a thick line.

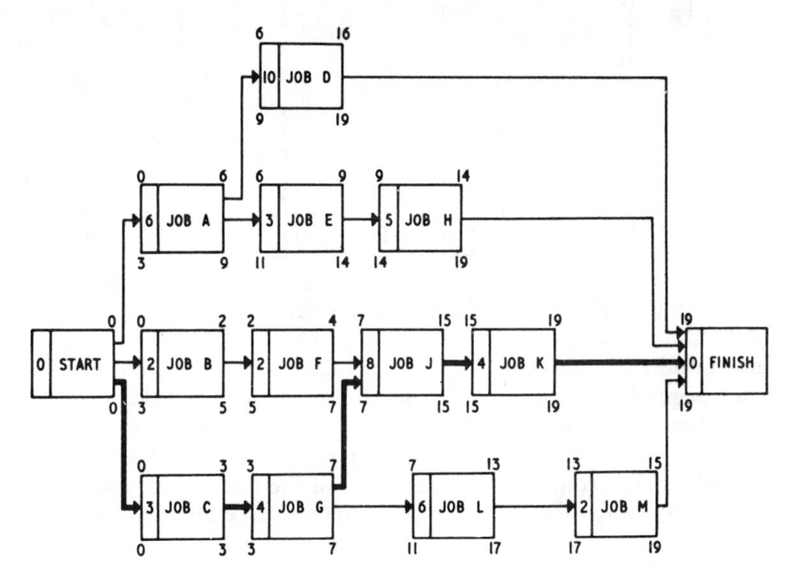

Fig. 4.13. Standard network—critical path shown by thick line

Another method sometimes used is to shade the critical job boxes instead of marking the arrows.

TABULAR PRESENTATION OF START AND FINISH TIMES

The network gives complete information about the start and finish times of each job. The people who are going to use this information may prefer to have it presented as a table, which is very simply constructed from the network and can take the form shown in Table 4.1. It is, of course, possible to arrange the jobs in any appropriate order, for example, by departmental responsibility or by putting all the critical jobs first.

Table 4.1. STANDARD NETWORK

Job No.	*Job Description*	*Duration (days)*	*Start*		*Finish*	
			Early	*Late*	*Early*	*Late*
1	A	6	0	3	6	9
3	B	2	0	3	2	5
5	C	3	0	0	3	3
7	D	10	6	9	16	19
9	E	3	6	11	9	14
11	F	2	2	5	4	7
13	G	4	3	3	7	7
15	H	5	9	14	14	19
17	J	8	7	7	15	15
19	K	4	15	15	19	19
21	L	6	7	11	13	17
23	M	2	13	17	15	19

Note that Table 4.1 is *not* a schedule because it does not give start and finish dates for the non-critical jobs. It simply gives the scheduling options. In the table, Job C *is* scheduled because it is critical. It must begin at the start of day 1 and finish at the end of day 3. The table however, does not schedule, for example, Job H. It simply says it *can* begin between days 9 and 14 and can finish between days 14 and 19. Someone has to *decide* when all the non-critical jobs will be run. The way these decisions are made is dealt with in Chapter 6.

THE ANALYSIS OF OVERLAPPING JOBS

In Chapter 2 the two methods of representing overlapping jobs were dealt with. Where jobs are split up into elements the analysis follows the lines already described. An example is shown in Fig. 4.14, which deals with the drawing, tracing and printing of a number of diagrams.

In the second method, lead and lag times are used, and the analysis is somewhat different. An example is shown in Fig. 4.15. This is the only occasion when times are placed against the arrows. All arrows with times on them represent 'leads' and 'lags' and they are drawn from the corners of the job boxes. Their meaning in Fig. 4.15 is as follows:

Arrow *a*: after 3 hours' drawing has been done tracing can start.
Arrow *b*: two hours' tracing is necessary after the completion of the last drawing.

Arrow *c*: after 2 hours' tracing has been done, printing can start.
Arrow *d*: one hour's printing is necessary after the last tracing has been completed.

DIFFERENCES IN THE METHODS OF SHOWING OVERLAPPING JOBS

The first method, where the jobs are split up into definite elements, is the most precise. The second method, sometimes called a 'ladder'

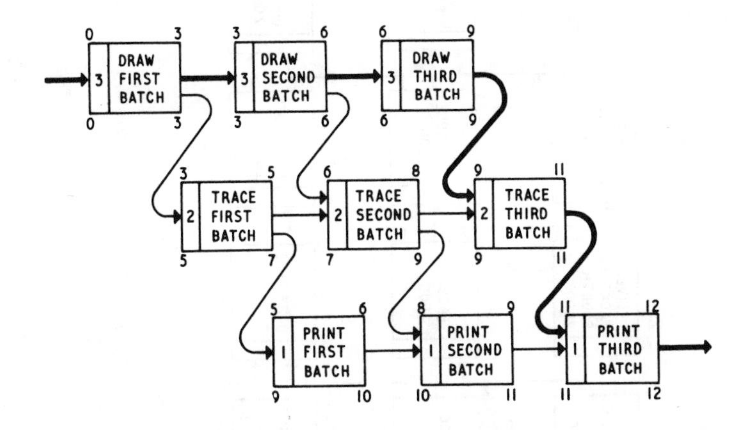

Fig. 4.14. Overlapping jobs—critical path shown by thick line (times in hours)

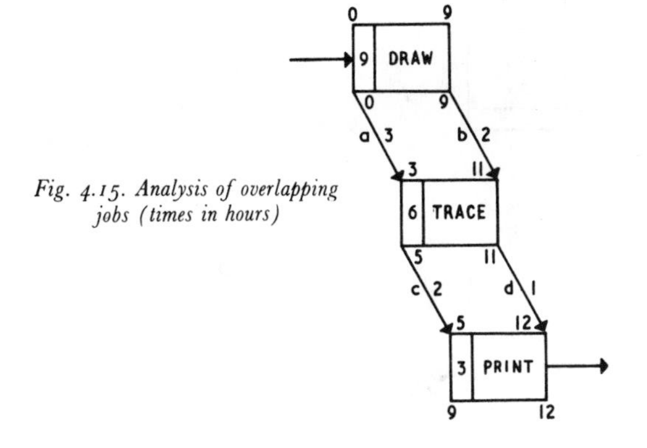

Fig. 4.15. Analysis of overlapping jobs (times in hours)

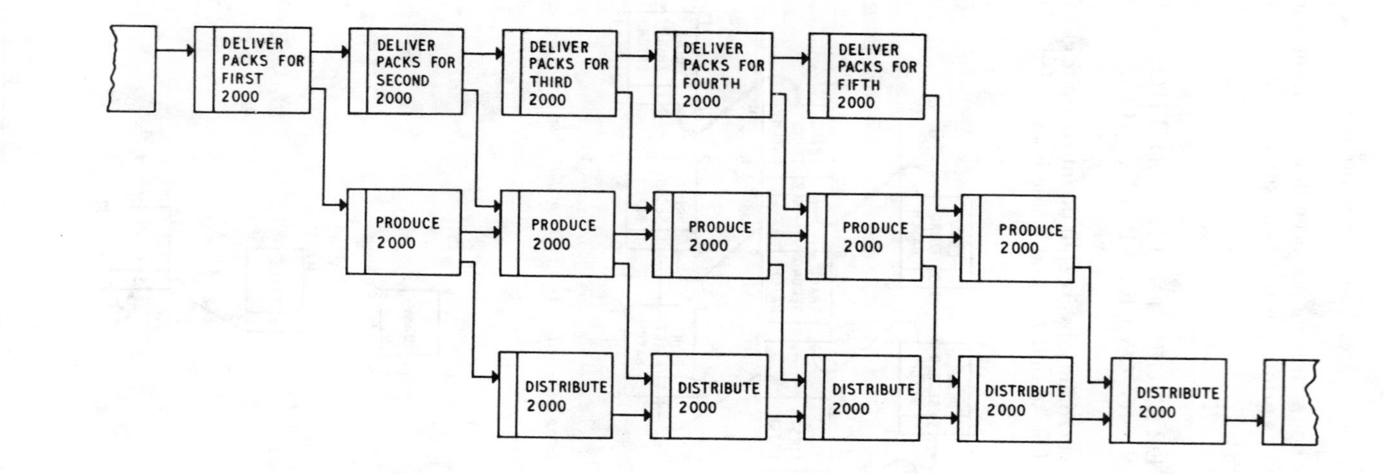

Fig. 4.16. Overlapping activities split into five elements

representation, is less precise and care has to be taken in the scheduling stage when ladders are used.

A PRODUCTION EXAMPLE

Figure 4.16 shows how overlapping activities of obtaining packing, production and distribution was handled in the launch of a new food product. The first method was used in this case, with a total production of 10,000 units broken down into 5 elements.

DETERMINATION OF 'SPARE' TIME AVAILABLE ON NON-CRITICAL JOBS

One term that ABC has borrowed from conventional network methods is 'float'. A job which is not critical has some time, over and above its duration, available for use if need be. It is this 'spare' time that is called float. Take an example from the standard network as shown in Fig. 4.17. The total time available for Job E is from the

Fig. 4.17

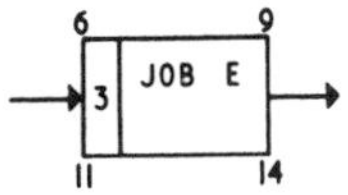

6		9
3	JOB E	
11		14

end of day 6, the earliest start, to the end of day 14, the latest finish, that is, 8 days. Since the job takes only 3 days there are 5 days to 'spare', or 5 days float. But the availability of this float to Job E depends entirely on when we decide to start it. If, for some reason, we schedule the job to begin on day 11 and end on day 14 there will be *no* float available to the manager who is responsible for this job. If, on the other hand, we decide that it will begin on day 6 and end on day 9, all 5 days will be available to him.

It is for this reason that the calculation of float is left until *after* the scheduling stage, as will be explained in Chapter 6.

SUMMARY

The method of showing start and finish times for each job has been given and the forward and backward pass calculations have been explained. The critical path and project duration is determined as a result of the forward and backward pass. A method of giving start and finish times in tabular form has been described. Overlapping jobs can be analysed by breakdown into elements or by ladders. Float is best determined after the scheduling stage. The network with earliest and latest times on the job boxes is the analysis bar chart.

5
Reducing the Duration of Projects

If the first analysis of the network produces a project duration that is too long, it will be necessary to reduce it. This will require a thorough understanding of the work involved in the project, but there are some general lines of approach that have proved useful.

Quite clearly, the only way to reduce the project time is to look in the first instance at the critical jobs, which usually number about a quarter of the total. As far as project duration is concerned it is useless reducing the time of the non-critical jobs; this fact in itself is an advance over less precise methods of planning. When the critical jobs have not been determined and a project is running late, it is quite common to find everybody working overtime. Usually, three-quarters of this effort has no effect on the completion date.

CHALLENGING THE METHOD

The most significant reductions in project duration usually arise from a change in the *method* of carrying out the project. First, it is necessary to establish that all the jobs in the critical path really do need to be there at all. This may seem obvious but it is surprising how often jobs can be eliminated altogether. There must be a satisfactory answer to the question: 'What is the purpose of this job?'

Striking examples have occurred in the overhaul of major equipment, where many jobs have been eliminated from the network dealing with the shut-down period and performed before or after the shut-down.

Another example of a change of method producing a reduction in project time arose with a company making electro-mechanical equipment. The original network critical path contained an element as shown in Fig. 5.1. The reason for this logic was that Assembly A was built up on the shop floor and Assembly B then constructed on top of A.

Challenging the method produced the question 'Why can't we build them at the same time, in parallel instead of in series?' The company produced a jig which represented Assembly A and built

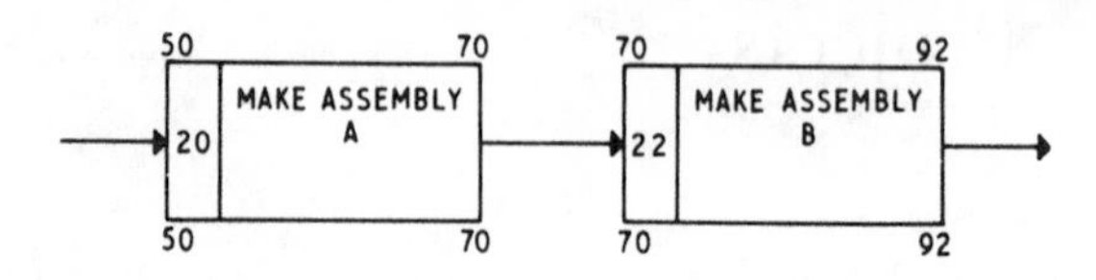

Fig. 5.1

the two assemblies side by side, mating them together when they were both completed. The network then became as shown in Fig. 5.2.

Of course, a check had to be made to ensure that resources would be available to carry out both jobs at the same time. This was so, and the critical path was reduced by 17 days at the cost of manufacture of the jig. The manufacture of A was also removed from the critical path as can be seen from Fig. 5.2.

A similar time-saving was achieved by the company who had to wire electronic equipment into trailers. Instead of waiting for the

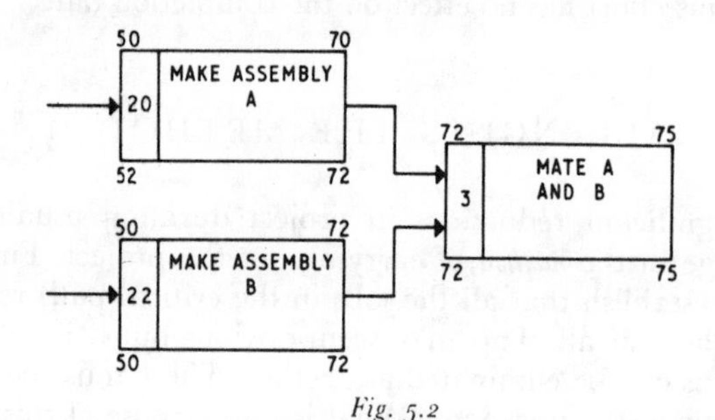

Fig. 5.2

trailers to arrive before commencing the wiring, cable looms were produced on jigs and installed partly completed.

These examples are just plain common sense, of course, something that any competent engineer could produce at the drop of a cloth cap. The point is that the available talent and time in the company is concentrated on the critical jobs, and the pay-off for ingenuity is maximised. There is little point in saving 5 days on a job that has 6 weeks float.

Sometimes, significant savings in time can be achieved by taking serial jobs and overlapping them. A marketing example is shown in Fig. 5.3. Originally, the plan for the production of packaging involved the submission of the proofs for all 16 varieties of the product at the same time. This was followed by approval of the proofs and

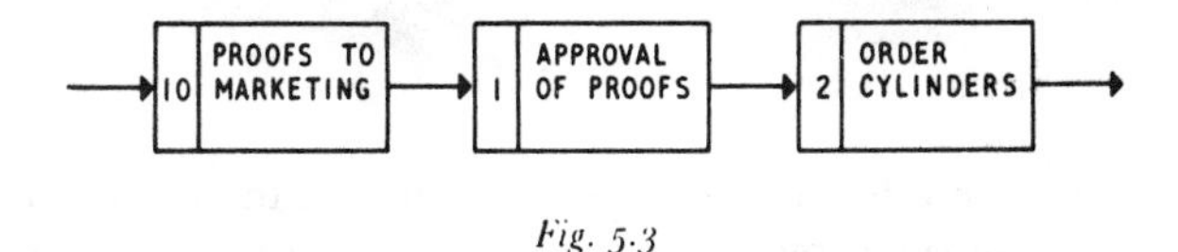

Fig. 5.3

the placing of the orders for the printing cylinders. This sequence was on the critical path and the logic was changed to that shown in Fig. 5.4, where the proofs were split into 2 groups. The time saved

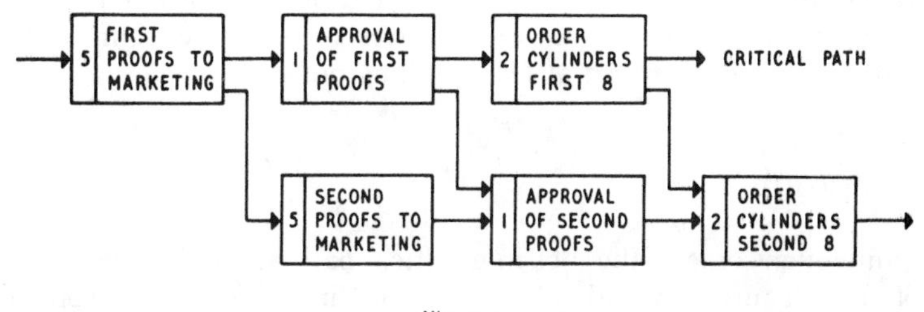

Fig. 5.4

was 5 days, since the second batch of proofs was no longer critical.

Again, just common sense, but the need to apply thought to this area did not become apparent until it was found to be critical.

REDUCING THE TIME OF INDIVIDUAL JOBS

When the possibilities of reducing the project time by changing the logic have been exhausted, and the project is still too long, it is necessary to look at individual jobs. The following factors are usually important when deciding which of the critical jobs to choose as candidates for shortening.

EARLY JOBS

Jobs which occur early in the project should be reduced before those that occur later. The first reason for this is that the whole project is brought forward. The second is that the option to reduce the later jobs remains open longer, and can be used as a kind of 'reserve' as the project takes place.

LONGEST JOBS

A long job often offers more scope for reduction than a shorter one, simply because the same percentage saving will produce a greater time-saving.

TECHNICAL DIFFICULTY

In a critical path some of the jobs will be easier to reduce than others for technical reasons. Again this involves knowing the technicalities of the work involved or, at least, finding out.

CONTROL

Quite often some of the jobs in a critical path are under the control of the organisation and others are not directly under its control. Typical examples are the delivery periods of suppliers and the times for sub-contracted jobs. Usually, the times for jobs which are under direct control can be reduced more readily than those that are not, but examples have occurred where the reverse was the case. The sub-contractors in these examples were brought to heel by use of a big stick, but everyone is not in the happy position of being able to use one.

COST OF REDUCTION

Some jobs are cheaper to reduce than others and where a choice exists the cheapest reduction is obviously preferable. This brings up an important point: it is not sufficient for management to be told that time can be 'saved' on a project; it also needs to know what the bill will be and how much will have to be paid for the time saved.

It is theoretically possible to 'buy' time on most projects provided there are no real technical constraints. The usual situation is that the cost is too high. We could employ another 10 draughtsmen next week *if* we paid them enough. We could install another 10 test bays *if* we could get the capital. We could fly the package designer to New York *if* someone would pay his expenses. Management has to judge how much it is prepared to pay for a time-saving if this is not free.

The network enables time–cost trade-off tables to be constructed which set out the options. The standard network is used as an example.

TIME–COST TRADE-OFF

The analysed network is shown again in Fig. 5.5. Assume that the

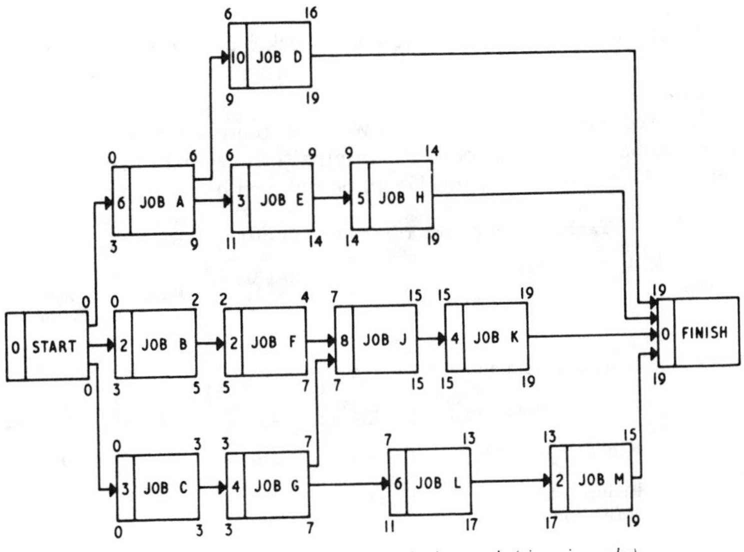

Fig. 5.5. Analysis bar chart for standard network (times in weeks)

time unit is now weeks. There are now 4 jobs on the critical path, C, G, J and K. Suppose that an examination of the factors listed earlier has led to the choice of Job C for the first reduction in time and that Job C can be reduced to 2 weeks at an extra cost of £100. The project will now be finished in 18 weeks.

Job G is the next job selected and this can be reduced to 3 weeks at an extra cost of £150, thus reducing the project to 17 weeks.

Suppose that the third job selected is J and that up to 4 weeks could be saved at an extra cost of £200 per week saved. It would be

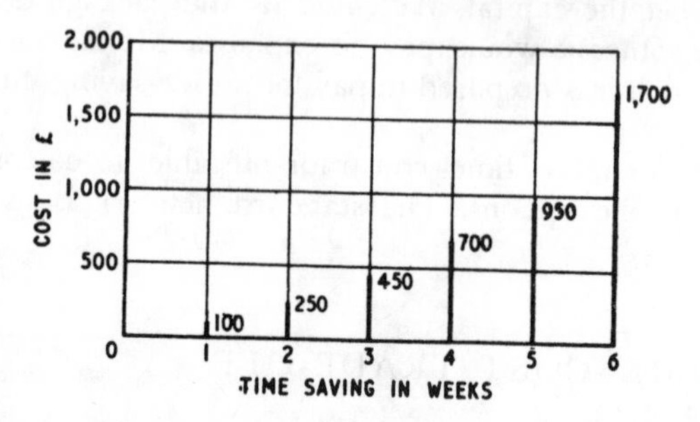

Fig. 5.6. Time–cost trade-off—standard network

little use saving more than one more week on the critical path because the job sequence A–D becomes critical when the project is reduced to 16 weeks.

If a reduction to less than 16 weeks is required, something will have to be done about both the original critical path and the new one, A–D. Suppose that Job D can be reduced by up to 3 weeks at a

Table 5.1. TIME–COST TRADE-OFF: STANDARD NETWORK

Step	*Action*	*Cost* (£)	*Cumulative Cost* (£)	*Weeks Saved*	*Project Duration*
1	Reduce C from 3 to 2	100	100	1	18
2	Reduce G from 4 to 3	150	250	2	17
3	Reduce J from 8 to 7	200	450	3	16
4	Reduce J from 7 to 6 *and* D from 10 to 9	200 50	700	4	15
5	Reduce J from 6 to 5 *and* D from 9 to 8	200 50	950	5	14
6	Reduce J from 5 to 4 *and* D from 8 to 7 *and* H from 5 to 4	200 50 500	1700	6	13

cost of £50 for each week. It would be useful to reduce D to 8 weeks (and J to 5) since this would reduce the project time to 14 weeks. At this point the sequence A, E, H becomes critical and further reduction in project time involves reducing one of these jobs.

Finally, suppose that Job H could be reduced by a week at a cost of £500. The project could then be reduced to 13 weeks, and all jobs except B and F will be critical.

The trade-off (Table 5.1) summarises the time that can be saved and the cost of this. The steps are sequential and the costs cumulative.

The results can also be presented graphically as shown in Fig. 5.6. It is tempting to join the points up, but this should not be done

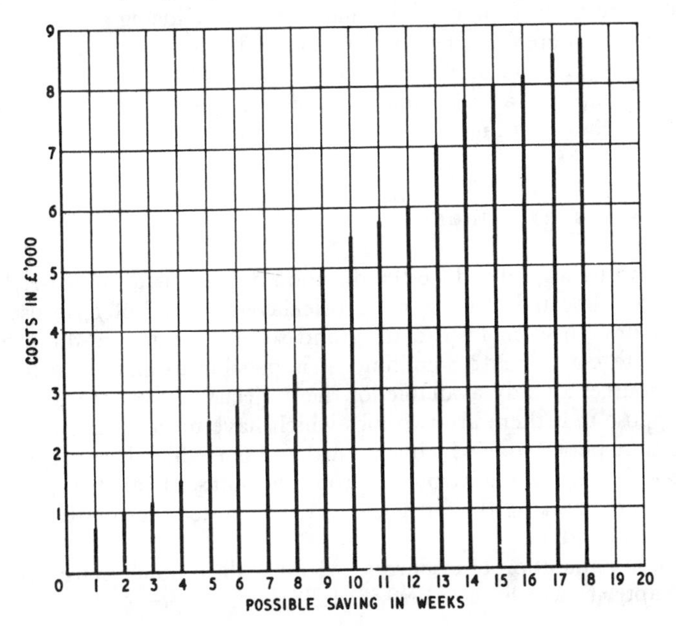

Fig. 5.7. Time–cost trade-off for Project PQ41. Network 65/A2/14 (costs from cost accounts)

because the time-savings between the points have not been examined.

Using this approach, management can find out not only how much time can be saved, but how much it will cost to save it. The options open have been quantified.

This simple example has been used to demonstrate the approach to problems of project time reduction. In this case nearly all the jobs became critical and it is possible in theory to reduce all projects in this way, so that all jobs are critical. If this is done it is very likely that, when the project starts, such a highly stressed plan will shatter as soon as any extra strain is placed on it.

How much flexibility to have in a plan is a matter for management judgement, and depends on the amount of uncertainty the project contains. If starting and finishing dates are fixed there is less room for manoeuvre.

Using the network to test alternative solutions is a form of 'simulation' and the network can be regarded as a 'model' of the project; so we are carrying out operational research after all!

In a real project there will be many ways of reducing project time and many different sequences to try out. The network helps to find the cheapest and most effective method of project time reduction.

An example of a cost–time trade-off table for a manufacturing project is shown in Fig. 5·7.

INCREASING RESOURCES

The obvious way to reduce the duration of a job is to increase the resources allocated to it: to use 2 bricklayers instead of 1; to use 3 design draughtsmen instead of 2, and so on. Where are these resources to come from? Sometimes it is possible to find them from the resources already available for the project.

Suppose that there are two jobs which have to take place in the same time period, one which is critical and one which is not. It may be possible to switch resources from the non-critical one to the critical one, making the former longer, which does not matter, and the latter shorter.

The manipulation of resources is dealt with in much more detail in Chapter 6, which describes scheduling.

SUMMARY

Step one in reducing project time is to challenge the method used. Step two is to reduce the duration of individual jobs. The network can be used to find the costs of saving time. To reduce the project time it is sometimes necessary to switch resources from non-critical to critical jobs.

6

Scheduling

Scheduling is the act of producing a timetable of work for the project, showing when each job is to begin and finish.

The critical jobs schedule themselves, as has been seen, but someone has to decide when all the non-critical jobs are to take place. The factors that have to be considered in making this decision will be examined later. The first step is to convert the analysis bar chart into a bar chart drawn against a time scale.

TIME-SCALE CHARTS

Consider Job L in the analysis bar chart, as shown in Fig. 6.1. The job must be scheduled somewhere between the end of day 7 and the end of day 17, which are the earliest start and latest finish, respectively.

There are many ways in which the job could be scheduled, three of which are shown. Schedule 1 can be called the earliest start schedule, in which the job is begun as early as possible and finishes as early as possible. The other extreme is Schedule 2, where the job begins and ends as late as possible. Schedule 3 is just one of the possibilities that exist between these limits.

However, it is not possible to schedule each job in isolation in this way, since there will usually be other jobs in the project that will be affected by whatever decision is made. The project has to be looked at as a whole.

The first step, therefore, is to draw the project against a time scale showing each job beginning as *early* as possible. This schedule will probably not be used finally but it is a good starting point. The standard network earliest start schedule is shown in Fig. 6.2.

Here, the critical jobs have been drawn so that they form a continuous bar from start to finish of the project. The other jobs have been shown beginning as early as possible and so all the float is at the end of each sequence, as shown by the striped bars in Fig. 6.2.

We are at liberty to 'slide' these non-critical jobs along their

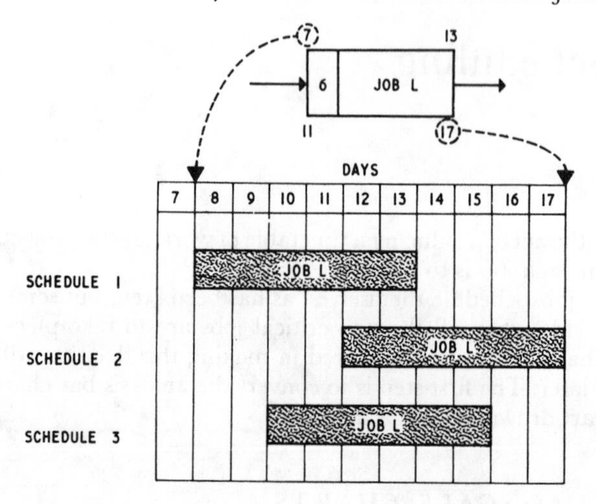

Fig. 6.1. Three possible schedules for Job L from the standard network

'slots' up to the limits of the float in each case.

Note that arrows are used to indicate how the jobs depend on each other. For example, L depends on G; J depends on F as well as on G.

FACTORS TO CONSIDER WHEN SCHEDULING

SIMPLE CASES

In some cases, making decisions about the scheduled start and finish dates is comparatively simple. For example, in a project concerned with the launching of a new product a network for use by senior management will show the major jobs to be carried out by the various departments concerned. There is unlikely to be any difficulty about the allocation of resources between departments since these are usually quite separate. In cases like this it is often decided to allocate float to jobs which might prove to be tricky. In one case, where a new production line was concerned, it was agreed that any float available would be allocated to production jobs.

RESOURCE ALLOCATION

A much more difficult situation arises when jobs in a project use common resources. This case is best shown by reference to the standard network. Suppose each job in the network involved a single type of resource, and let this be called 'men'. In deciding the job durations we said that these should be estimated on the basis that a 'normal' level of resource would be used. In Fig. 6.3 the number of men 'normally' required to carry out each job is shown against the bar representing the job.

Under the bar chart the number of men required on every day of the project has been entered. This figure is arrived at by totalling the men required on all the jobs scheduled to be run each day. The

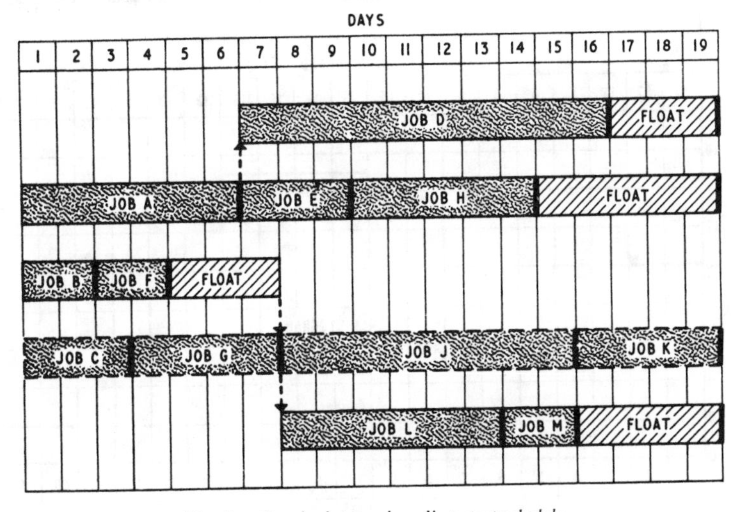

Fig. 6.2. Standard network earliest start schedule

requirement can be expressed graphically and this method of representation is also shown in Fig. 6.3 as a histogram.

It is clear from the diagram that as the jobs with float are moved between their scheduling limits, the pattern of men required will change also.

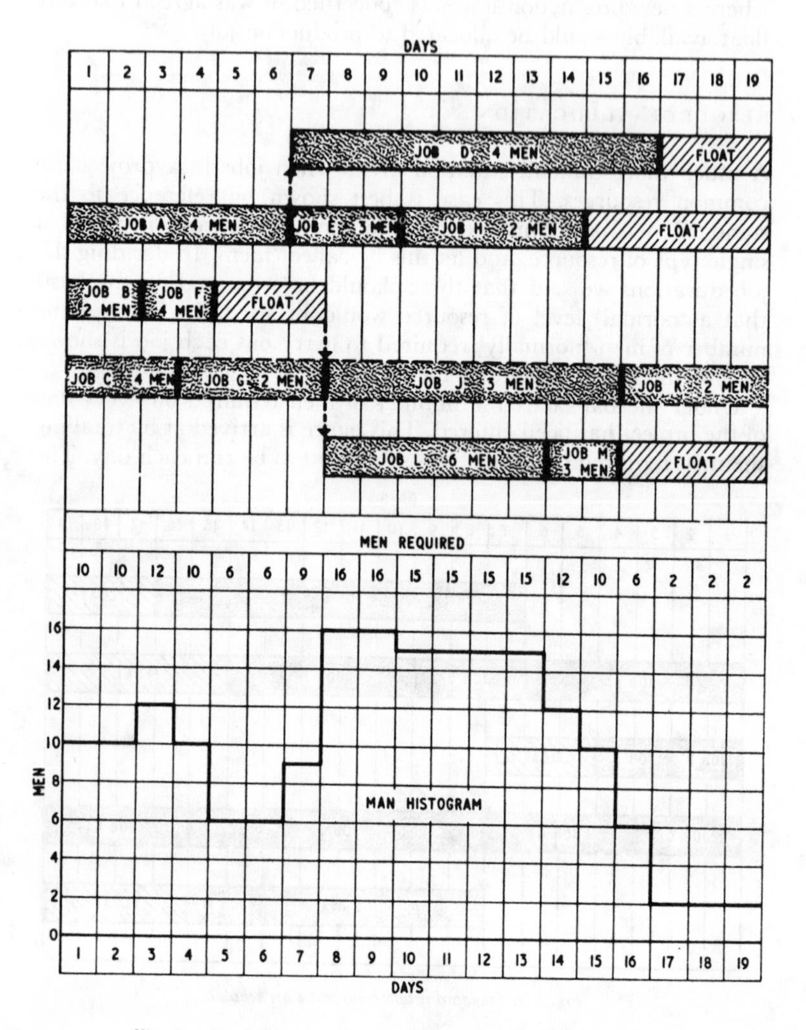

Fig. 6.3. Standard network—men required for earliest start schedule

In Fig. 6.4 all the jobs have been scheduled as late as possible, with all the float placed at the beginning. The pattern of labour usage has changed to that shown. If this schedule were used all the jobs would be critical.

To look at all the schedules possible between these two limits would take a very long time as there are an extremely large number of arrangements that could be made, even in this simple project. (Readers who enjoy puzzles might like to work out exactly how many unique schedules can be constructed.)

Like all sequencing problems this one is very difficult to solve. Let us agree that we want the 'best' schedule. What is 'best'? If it is the one which uses men in as 'even' a way as possible the ideal

Table 6.1. CALCULATING THE 'EVEN' USE OF MEN IN A SCHEDULE

Job	*Duration*	*Men Required*	*Work Content (man-days)*
A	6	4	24
B	2	2	4
C	3	4	12
D	10	4	40
E	3	3	9
F	2	4	8
G	4	2	8
H	5	2	10
J	8	3	24
K	4	2	8
L	6	6	36
M	2	3	6
			189 *total*

189÷19=10 men per day, except on one day when 9 will be required.

schedule would result in the histogram shown in Fig. 6.5. This has not been obtained by manipulating float but by working out the number of man-days of work and dividing by the number of days available, as shown in Table 6.1.

Unfortunately, this histogram is not possible because of the logic restraints in the project. The best that the author has been able to do is shown in Fig. 6.6 (page 54). Better solutions may, of course, be possible. However, if one is not careful it is possible to spend more time scheduling than doing the job! Note that the decision to run jobs D and H as late as possible makes them critical also.

Figure 6.7 (page 55) compares the histograms for earliest and latest starts with the author's best schedule. Some improvement has been achieved, if improvement is defined as 'labour smoothing', since the maximum number of men needed on any day is 13.

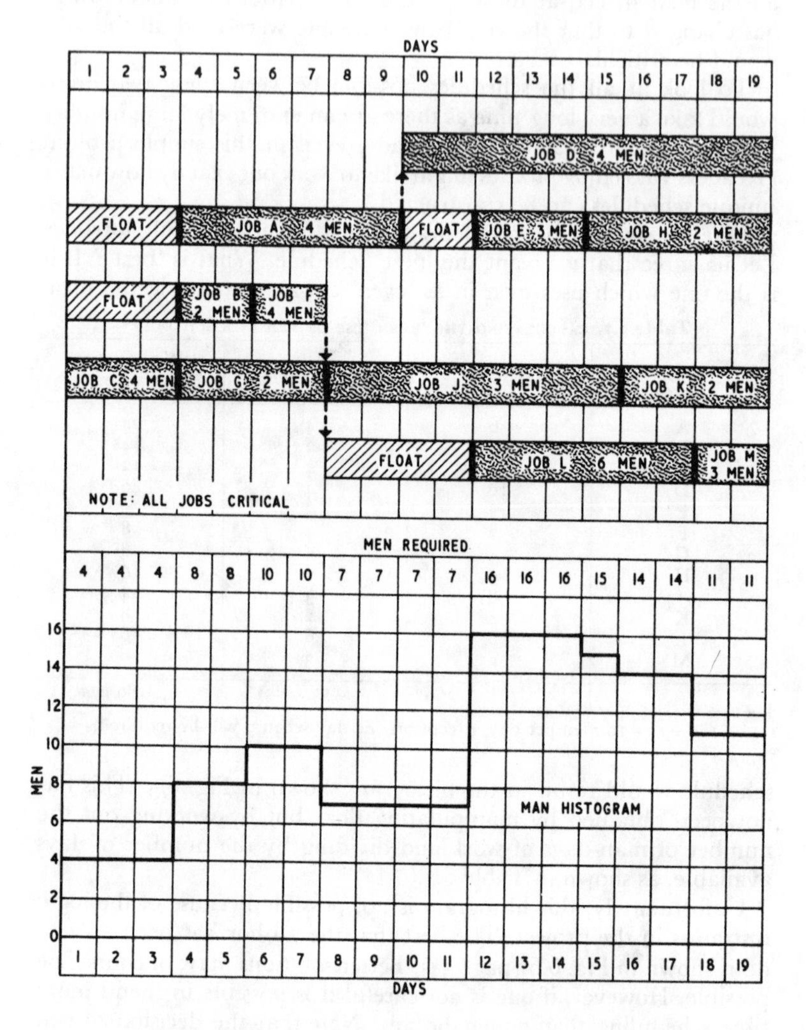

Fig. 6.4. Standard network—men required for latest start schedule

MATCHING RESOURCES REQUIRED TO RESOURCES AVAILABLE

A more usual problem is not to have to look for an optimum solution as just described, but to try to match what we need with what we have.

Using the standard network again, assume that there are only 13 men available at most. Redrawing the histograms for earliest start

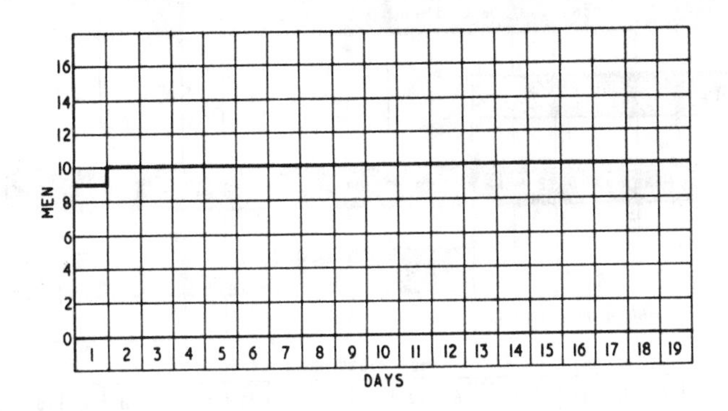

Fig. 6.5. Standard network—theoretical 'best' use of labour

and latest start schedules, with the resource limit shown also, results in Fig. 6.8 (a) and (b), page 57, where the overload on days 8–13 and 12–17, respectively, can be seen. The 'best' schedule overcomes this difficulty as shown in Fig. 6.8 (c).

RESOURCE-LIMITED CASES

What happens if the assumption made during the logic stage about unlimited resources turns out to be untrue? Suppose, for example, that in the standard network there are only 10 men available at maximum. The project must now be re-scheduled so that this limit is not exceeded. Figure 6.9 (page 59) shows one solution to this problem and there are many more, some possibly better. The project time has had to be extended, not because the logic or time of individual jobs were wrong, but because resources are not available at the necessary level; hence the term 'resource-limited schedule'.

Note that the float shown in Fig. 6.9 can be used only if men are

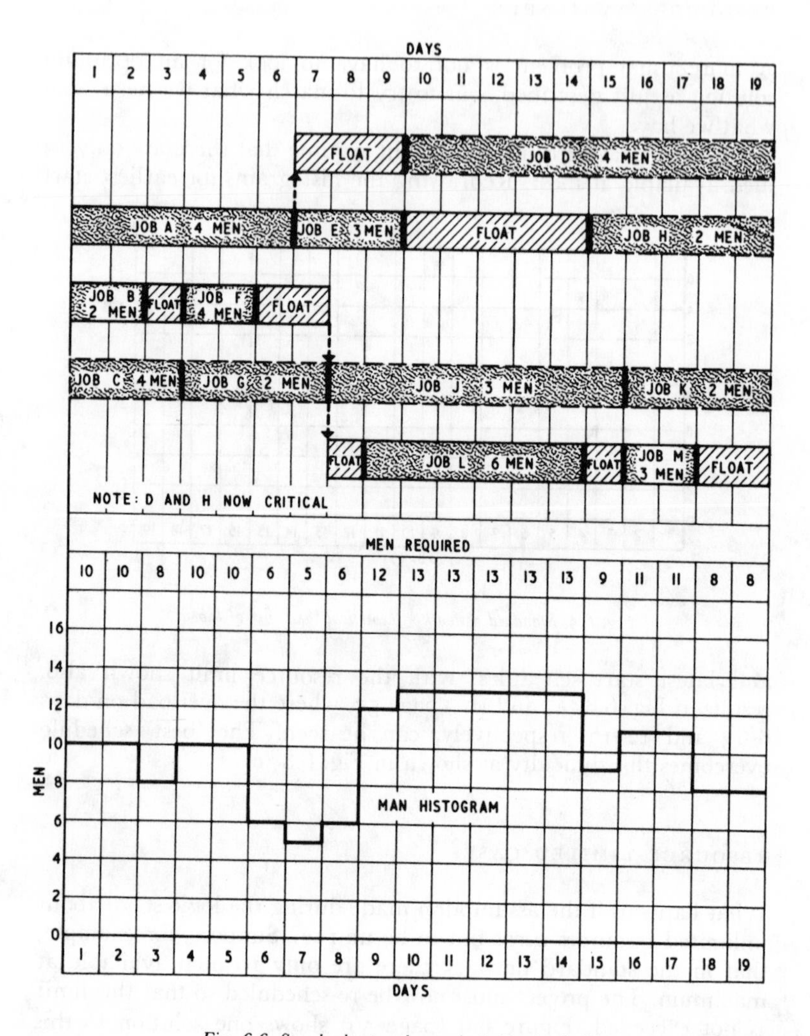

Fig. 6.6. Standard network—author's 'best' schedule

available to carry out the job. For example, Job B could be extended or delayed by 2 days without causing an overload, and Job F could be extended by 1 day. These extensions or delays are not independent of each other, however, and their relationship can be seen very clearly by examining the bar chart and the histogram. Each job must be examined to find out if it is critical from the point of view of *time* and *manpower*.

If, on the other hand, it is mandatory to finish in 19 days, the extra resources will have to be found from somewhere. Since we

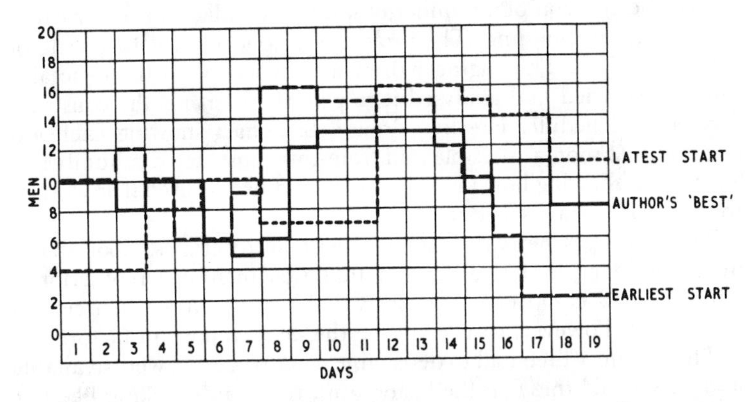

Fig. 6.7. Standard network—three labour histograms compared

know exactly how many more will be needed, and when, the additional costs can be calculated. This type of schedule is known as a 'time-limited schedule'.

If analysis of resources shows that there are not enough, either more have to be obtained or the project time has to be extended, or a combination of both. There is no avoiding this decision. The instruction to 'manage with what you have *and* get it done in 3 weeks' is an unrealistic one, if the calculations are correct. If such instructions are given, whoever gives them is saying the planner has been wasting his time. We are back to 'head down and charge' methods.

MULTIPLE RESOURCES

When jobs use more than one resource, for example, men and machines, the difficulties of scheduling increase. Here we have to try to minimise the use of several resources at the same time, or not to exceed several resource limits at the same time.

Unfortunately, there is only one kind of float, not one float for each type of resource, and using the float to smooth the use of resource *A* will probably unbalance some of the others. It becomes necessary to list the resources in order of priority and to tackle the scheduling as a sequence. The procedure is best illustrated by an example.

An engineer was concerned with managing a project which involved the erection of a number of large assemblies of prefabricated metal in a test building. The project was concerned with testing of aircraft engines. The engineer drew a network to get the logic and timings recorded and analysed it to give the basis for the construction of his schedule. He then drew a bar chart showing each job beginning as soon as possible and from this constructed a number of histograms showing how his resources would be required if he worked to the earliest start schedule.

The resources he was concerned with were cranes, floor space, fitters, welders, and riggers, and that was their order of priority which in this case depended on availability of extra resources. In other cases, however, costs could be the most important factor.

The engineer then had to determine what resources were available and he showed thesa on the histograms, the results looking like Fig. 6.10 (page 60). It will be seen that 'overloads' existed for every resource if the earliest start schedule were used. In this case, a man 'overload' meant more tradesmen than the engineer had available. In the case of cranes it was impossible to get more than three into the building. Space had been allocated to him by the overall project management.

Starting with cranes, he used the float available on the non-critical jobs that required cranes and easily constructed a schedule that needed no more than three cranes at any time. Space was more of a problem and he was unable to manipulate starting dates to avoid an overload altogether. The best he could achieve was to limit the overload to one week. Armed with his charts he approached the project management and succeeded in securing an extra allocation during that week. (They thanked him for giving them three months' notice and asked him to lecture to their project managers' training course on 'Planning by Networks'!)

Having sorted out his two main resources he set to work on his tradesmen and was able to reduce the peaks and troughs considerably. No extra fitters or welders were required, but four extra

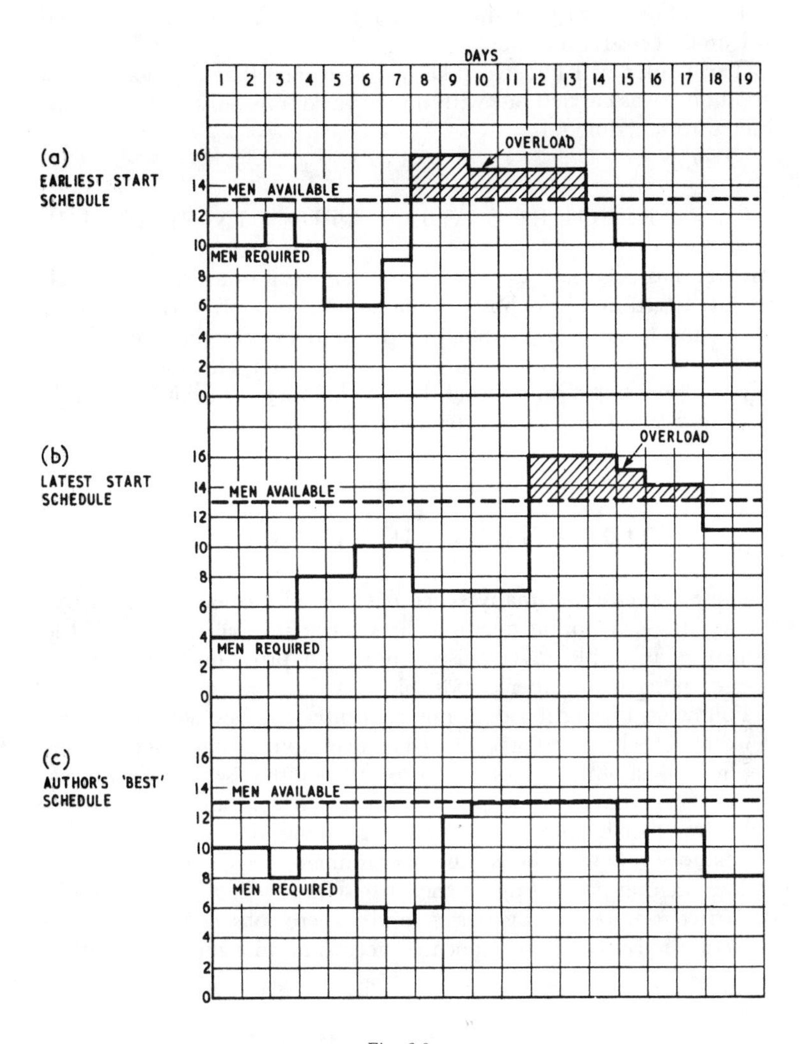
DAYS
1
2
3
4
5
6
7
8
9
10
11
12
13
14
15
16
17
18
19
(a)
EARLIEST START
SCHEDULE
OVERLOAD
MEN AVAILABLE
MEN REQUIRED
16
14
12
10
8
6
4
2
0
(b)
LATEST START
SCHEDULE
OVERLOAD
MEN AVAILABLE
MEN REQUIRED
16
14
12
10
8
6
4
2
0
(c)
AUTHOR'S 'BEST'
SCHEDULE
MEN AVAILABLE
MEN REQUIRED
16
14
12
10
8
6
4
2
0

Fig. 6.8

riggers were needed at the beginning of the project and these he was able to obtain, again mainly because he was able to give advanced warning of his requirement.

The reader will have to take my assurance that the plan was a good one, was carried out with minor variations only and was completed to schedule.

RESOURCE ALLOCATION IN PRACTICE

Enough has been said to show the principles of using float to match resources required to those available. A word of warning: don't go on too long trying to make better and better schedules. Don't forget that the work has to start. When you get a schedule that looks reasonable—stop. Don't forget that the number of schedules possible in even a small project is astronomical, something of the order of 10^{10}.

'RULES' FOR SCHEDULING

Because there are so many schedules possible, rules or procedures for making scheduling decisions have been produced in much the same way as people have devised 'rules' and 'procedures' for playing chess, where the number of possible moves is also very large.

Following these rules does not guarantee the best solution (even if we know what 'best' is) but it has been found that a set of rules applied consistently gives better results than a series of inspired guesses. (Anyone who is trying to 'teach' a computer to schedule must devise such rules. That is why it is taking so long to get a really satisfactory resource allocation programme.)

The starting point is again the earliest start schedule. Scheduling is carried out period by period and, where jobs compete for resources, the resources are allocated according to a number of rules, e.g.:

1. Give the resource to the job that has least float.
2. If jobs have equal float give the resource to the longest job.
3. Where there is conflict between jobs, give the resource to the job that uses the largest amount of the resource.
4. Where there is more than one overload in a period, deal with the one that involves the resource of highest priority first.

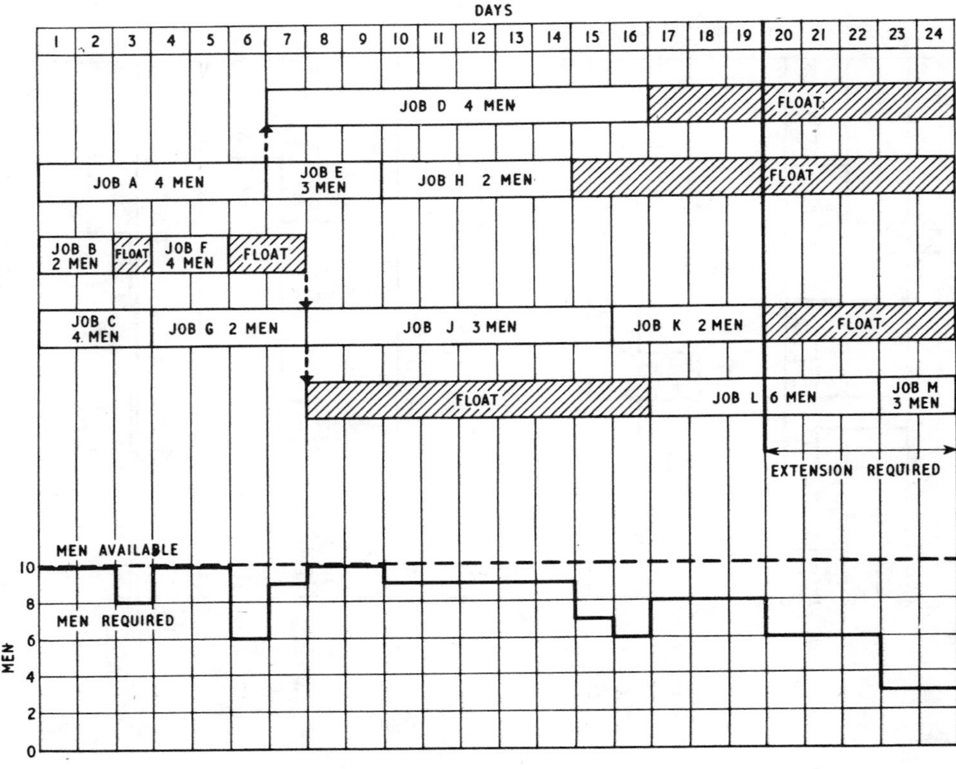

Fig. 6.9. Standard network—possible schedule with resource limitation of 10 men

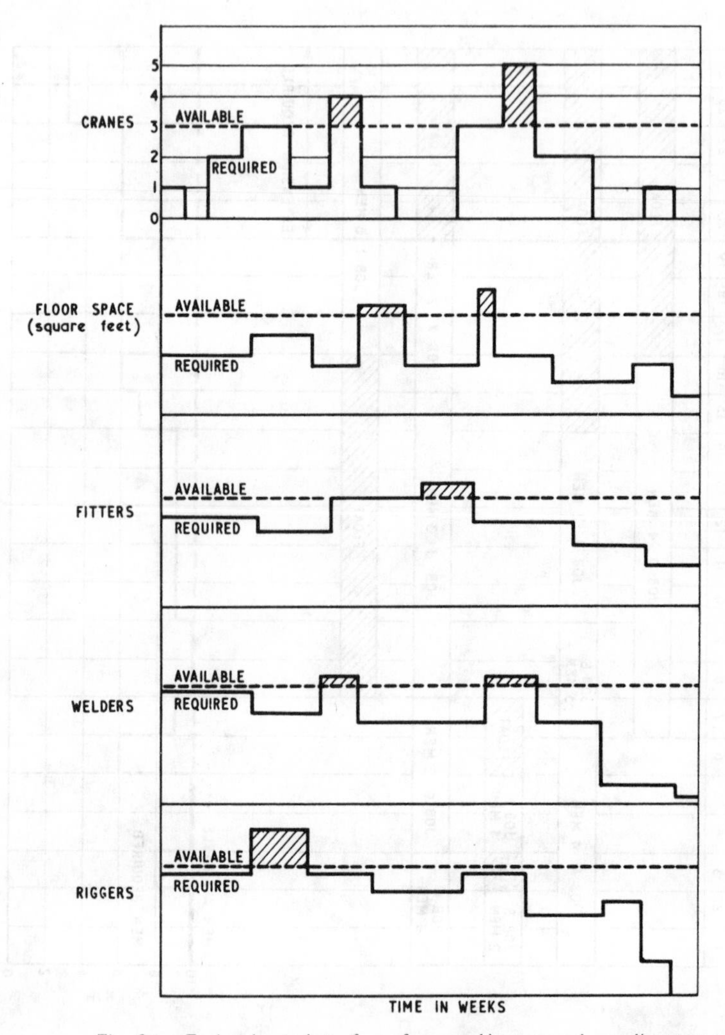

Fig. 6.10. Engineering project—form of resource histograms when earliest start schedule is constructed

The important point is to produce such a set of rules and stick to it.

Let us try this approach on the standard network, considering just the one resource, 'men', with a limitation of 12 men. Figure

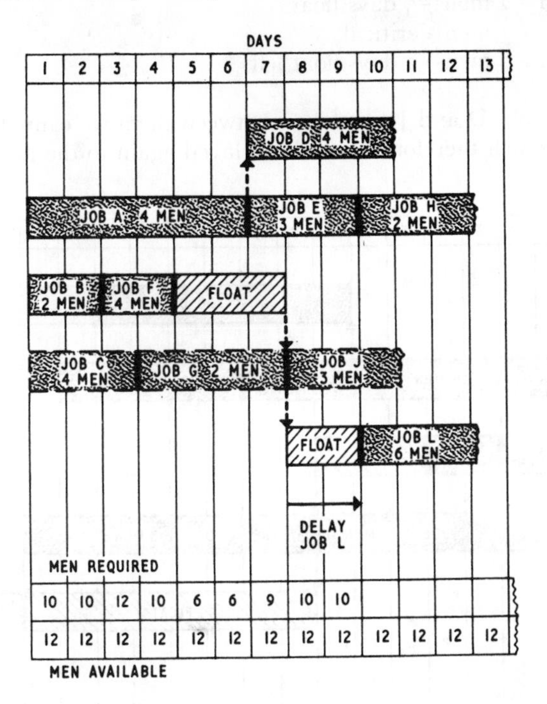

Fig. 6.11(a). Standard network—first stage of schedule construction

6.11(a) shows that enough men are available until day 8 when the following jobs are competing for men:

Job D—4 men—started on day 7.
Job E—3 men—started on day 7.
Job J—3 men—critical.
Job L—6 men—has 4 days float.

If we assume that a job once started must be continued, we are already using 7 men. Job J is critical and we must schedule this, bringing the total to 10 men. Job L must therefore be delayed at least until the next change occurs. This is at day 10.

When we get to day 10 the following jobs are competing for men:

Job D—4 men—must be continued.
Job H—2 men—5 days float.
Job J—3 men—critical.
Job L—6 men—2 days float left.

Since jobs D and J use 7 men between them we cannot schedule Job L, which therefore has to be delayed again to the next change,

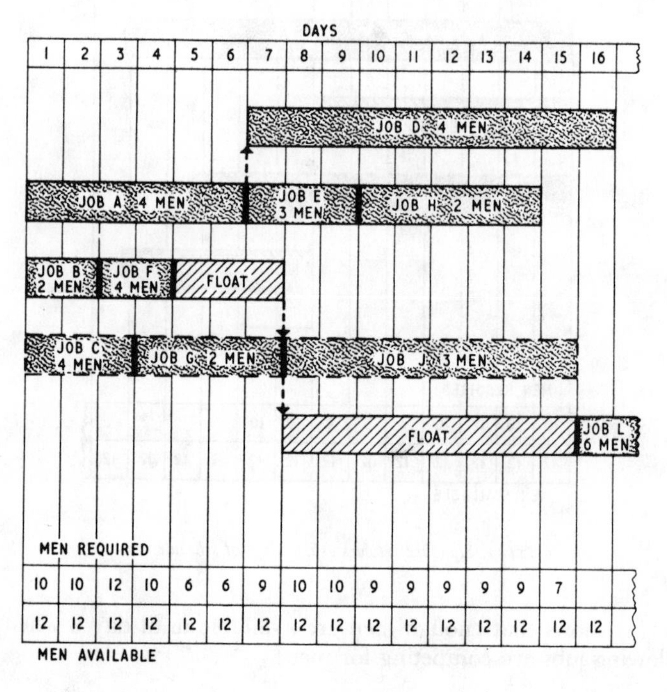

Fig. 6.11(b). Standard network—second stage of schedule construction

at day 15. This situation is shown in Fig. 6.11 (b). However, we can schedule Job H.

Even at day 15, Job L cannot be scheduled, since Jobs D and J require 7 men between them. Job L will have to be delayed until day 16.

Finally, the schedule becomes as shown in Fig. 6.11(c). Note that Job F has been delayed to reduce the manpower required on day

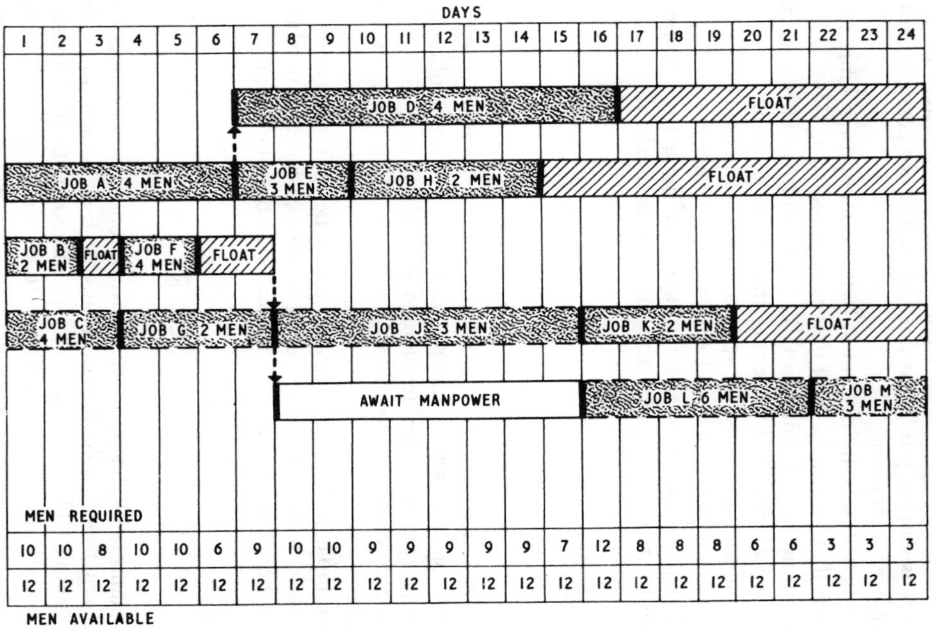

DAYS

JOB D 4 MEN
FLOAT
JOB A 4 MEN
JOB E 3 MEN
JOB H 2 MEN
FLOAT
JOB B 2 MEN
FLOAT
JOB F 4 MEN
FLOAT
JOB C 4 MEN
JOB G 2 MEN
JOB J 3 MEN
JOB K 2 MEN
FLOAT
AWAIT MANPOWER
JOB L 6 MEN
JOB M 3 MEN

DAYS	1	2	3	4	5	6	7	8	9	10	11	12	13	14	15	16	17	18	19	20	21	22	23	24
MEN REQUIRED	10	10	8	10	10	6	9	10	10	9	9	9	9	9	7	12	8	8	8	6	6	3	3	3
MEN AVAILABLE	12	12	12	12	12	12	12	12	12	12	12	12	12	12	12	12	12	12	12	12	12	12	12	12

Fig. 6.11(c). Standard network—final schedule

3 to 8 men. It can also be seen that by delaying Job K by 1 day, the manpower requirement on day 16 can be reduced to 10. Job K is no longer critical.

Note also that, although the logic of the network states that Job L can begin as soon as Job G has finished, this is not possible because of manpower limitations.

Applying the scheduling 'rules', with a manpower limit of 12, has shown that an extension of time is required to 24 days, but that 12 men are no more useful than 10. This result could hardly have been arrived at intuitively, even on this simple example.

The example illustrates the approach used to produce schedules. Graph paper and a pencil and eraser are the only essential equipment. The job is made easier by using one of the many proprietary planning boards that are available, and by an inexhaustible supply of patience.

SPLITTING JOBS

We have assumed so far that jobs, once started, must be continued. This may not be so and much greater flexibility can be obtained if jobs can be split up. The decision to split a job with float spaced between the elements requires knowledge of the technicalities involved.

For example, a six-day design job might be split into two groups of three days, with a gap of a week in between, whereas a machining job of six hours might have to be run as a continuous operation.

REDUCING RESOURCES

In the example of the standard network, on day 15 we made the decision not to schedule Job L because it took 6 men and only 5 were available. This is computer-type thinking. Being human we can ask, and answer, the question: 'Can we start Job L with 5 men and extend the time a little?'. If the answer is 'Yes', then we can go ahead and schedule Job L.

The difference between scheduling manually and scheduling by computer is that the former is intelligent and the latter mechanical. If only we could do sums as quickly as computers we would be well away. (Perhaps a better solution would be to produce an intelligent computer!)

CASH AS A RESOURCE

Latest start schedules are popular with accountants because, if starts can be delayed, payments can be delayed also. It is possible to construct cash flow diagrams from the schedule showing the improvement in cash flows that would result if everything were done as late as possible. One major international group will authorise capital expenditure on projects only when this is supported by a network showing when the cash will be needed. This is in addition to the normal financial justification required.

FLOAT

Readers familiar with conventional network methods will be acquainted with total and free float and other classifications, too. In ABC, float is calculated *after* the scheduling decisions are made, from the bar chart.

What the manager is interested in is the float actually available in practice. For this purpose the scheduling decisions must be considered and the float calculated on the basis of scheduled dates.

With visual display on a bar chart it is hardly necessary to work out float at all, as long as the scheduling limits are clearly defined.

After the schedule has been constructed, giving start and finish dates for each job, the float available can be seen on the bar chart. It is useful to work out this float because it will be used during the control phase, when the work is under way.

Float was mentioned briefly in Chapter 4. It is the measure of spare time on each job. Take the 'best' schedule produced for the standard network as shown in Fig. 6.6. The tabular presentation of start and finish times (see Table 4.1) can now be amended as shown in Table 6.2.

In addition to early and late start and finish dates the scheduled dates have now been added—we have made our decisions. In practice, it is normal only to issue the scheduled dates to those men who have to do the jobs. Note that float has been classified under two headings: 'total' and 'free'.

Total float is the time by which a job can be expanded or delayed without making the project late. If it is used up it may involve re-scheduling a subsequent job.

Free float is the time by which a job can be expanded or delayed without affecting a subsequent job.

Table 6.2. TABULAR SCHEDULE DERIVED FROM FIGURE 6.6

Job No.	*Job Description*	*Duration (days)*	*Men Required*	*Start*			*Finish*			*Float**	
				Early	*Scheduled*	*Late*	*Early*	*Scheduled*	*Late*	*Total*	*Free*
1	A	6	4	0	0	3	6	6	9	3	0
3	B	2	2	0	0	3	2	2	5	3	1
5	C	3	4	0	0	0	3	3	3	0	0
7	D	10	4	6	9	9	16	19	19	0	0
9	E	3	3	6	6	11	9	9	14	5	5
11	F	2	4	2	3	5	4	5	7	2	2
13	G	4	2	3	3	3	7	7	7	0	0
15	H	5	2	9	14	14	14	19	19	0	0
17	J	8	3	7	7	7	15	15	15	0	0
19	K	4	2	15	15	15	19	19	19	0	0
21	L	6	6	7	8	11	13	14	17	3	1
23	M	2	3	13	15	17	15	17	19	2	2

* Floats are derived from the bar chart; they result from the scheduling decisions. Before floats can be used manpower availability must be checked.

If jobs use up their float, this is equivalent to re-scheduling the project. The resource requirements of such a re-scheduling must be examined to see if resources are still available.

Take, as an example float, Job B. It can expand by one day (day 3) without affecting Job F, which follows it. It can expand by 3 days (days 3, 4 and 5) if we are prepared to delay the start of F until the beginning of day 6. It, therefore, has 1 day free float and 3 days total float. However, none of this float may be available if resources are so limited that the new schedule is not possible.

Note that in the example some jobs that were not critical in the analysis stage have been made critical by the scheduling decisions. In a latest start schedule all the jobs are critical. It is interesting to note that a latest start schedule is what people intuitively work to when they are using 'common sense'. The question usually asked about jobs is 'when must it be finished by?' The answer leads to the latest start. Under these conditions it is not surprising that things go wrong. ABC determines 'how soon' as well as 'how late' and then goes on to answer the question 'when is the best time?'

In real projects, day scales can be replaced by calendar date scales. As long as allowances are made for holidays, etc., no difficulties arise during the conversion.

SUMMARY

Scheduling is an application of ingenuity and common sense within a framework of the determined start and finish dates. When resources are concerned there is no optimum solution available but a good guide to a satisfactory deployment of resources can be obtained.

7
Control

The first six chapters of this book dealt with the planning stages. If the steps have all been carried out correctly, we now have a realistic plan with all the required resources neatly lined up against the scheduled start and finish dates for each job in the project. Someone presses the button, the project is launched—and the trouble starts.

ATTITUDE TO CHANGES

Anyone who finishes a project without having had to amend the network and bar charts several times is either a genius, very lucky, or has not updated the network to show what has actually happened.

Of course, changes will occur; how can anyone plan with absolute certainty? Some jobs will take longer than was estimated, some will be finished more quickly. Snags will arise which may lead to a change in logic. External changes may be impressed on the project management.

One reaction to these changes is for the man or men who did the planning to sit down and cry. 'Oh dear', they will say, 'our lovely plan has been spoilt by the harsh real world; our network has been ruined.'

One can understand this reaction but not sympathise with it. Networks help management to plan logically and quickly using the information currently available. They express management's intentions to execute the project in a given way. Management is entitled to change its mind when conditions change; indeed, quite often it must.

The network and the schedule must be regarded as a means of displaying the work situation as it actually exists. Networks do not create situations, they display them. The correct reaction to change is pride in the rapid and accurate way the network is used to represent the changes that have taken place. 'Look', the planners should say, 'as soon as something happens "Out There" our network reacts: it is like a living thing that adapts and changes itself in step with the harsh real world.'

Unfortunately, a self-adapting network has not yet been devised. A great deal of thought has to go into designing fast-reacting and

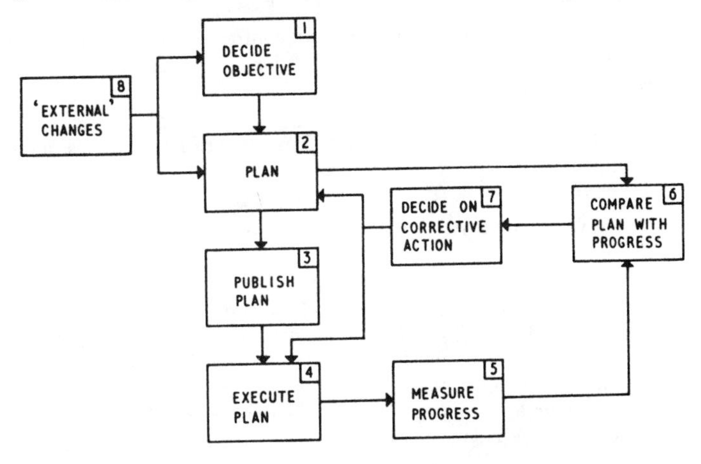

Fig. 7.1. Control cycle

efficient control systems. This chapter is concerned with some of the basic design considerations and shows how ABC can be used to control the execution of the plans that it has helped to make.

THE CONTROL CYCLE

Figure 7.1 represents the control cycle that operates in project management.

Box 1. An objective is the prerequisite to any plan.

Box 2. Production of the plan to achieve the objective. The way ABC can be used to produce a plan has been discussed.

Box 3. Publication of the plan. The best plan is useless unless the people who have to operate it understand their part of the plan and accept that it is valid. ABC helps here in a number of ways. First, it was stated earlier that the logic and timing stages should involve the people who are concerned. People who have helped to produce a plan are 'in the picture' from the outset. Secondly, the analysis bar chart and the schedule produced from it, form a very useful means of publishing the plan. Bar charts can be read and understood very much more quickly than a series of written instructions.

Box 4. The execution of the plan is a matter for line management. ABC helps because what has to be done is set down precisely against a time scale.

Box 5. Measurement of progress. If control is to be exercised it is necessary to measure progress in order to find out how much has been accomplished.

There are two important design considerations here, as set out below:

FREQUENCY OF MEASUREMENT

It is necessary to decide how frequently to measure progress on the various jobs in the project. Certainly, the start and completion of jobs need to be reported and it may be necessary to determine progress between these dates. The length of the project is a factor to be considered. In a project that is scheduled to run for a year, monthly measurements of progress may be sufficient for control at senior management level. Weekly measurements may be needed at a more junior level.

The principle to follow should be that measurements are taken as infrequently as possible. Why collect data that is not going to be used?

ABC gives a good indication of where the available measurement effort should be applied. It is obviously more important to measure progress on the jobs that are critical, or near-critical, than on the jobs that have plenty of float.

As a general rule, reports of achievement should be in the form of how much is still to be done, rather than how much has been done. This enables the updating of the network to be carried out more easily.

ACCURACY OF MEASUREMENT

It has become a cliché that control systems are no better than the data they handle; nevertheless it is true. An updated network is supposed to represent the situation that exists. If reports of progress are inaccurate the network will not represent the situation. The fault lies in the inaccuracy of the data, not in the control system. Managers who are in charge of jobs in the project must be convinced of the need for accuracy. There should be no 'fudging' of the data fed in. We need to know what has happened, not what we would have liked to have happened.

Box 6. Comparison. At this stage in the cycle it is necessary to compare progress with plans. This can be done very easily on the schedule bar chart. An example is shown in Fig. 7.2, where progress

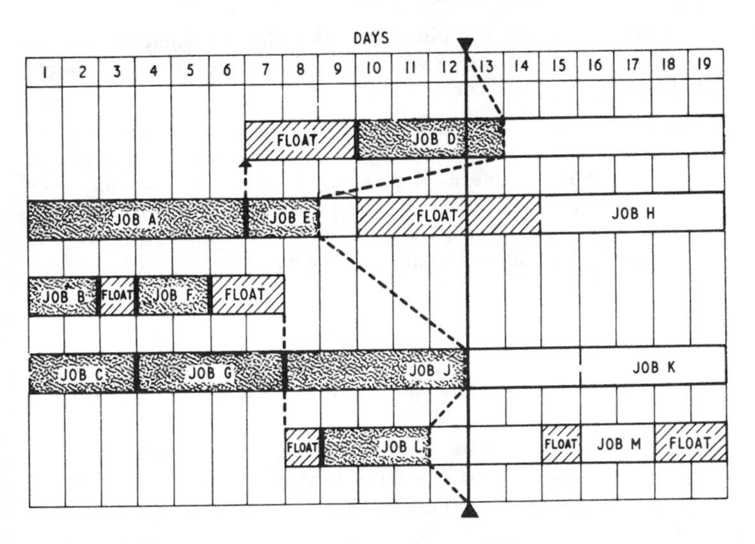

Fig. 7.2. Displaying progress and plan

on the standard network is shown. Assume we are at day 12. The cursor indicates this and the broken line joins the points of achievement of the individual jobs, which are also shaded to show how much has been achieved.

It can be seen quite clearly that:

Job D is one day ahead of schedule;
Job E is one day behind schedule;
Job J is on schedule;
Job L is one day behind schedule.

Box 7. Deciding on corrective action. Corrective action can take two forms, an alteration in execution to put the performance back in accordance with plan, or an alteration to the plan if the original plan is no longer possible. In either case what has to be done to correct the situation can be seen very easily by reference to the schedule bar chart. In the situation displayed in Fig. 7.2 the corrective action necessary might be as follows:

Job D: No action necessary.

Job E: No action necessary as long as the job can be finished in time for Job H to start on day 15, since E has 5 days free float.

Job L: No action necessary as long as L can be finished by the end of day 17 at the latest. L has 1 day free float, M has 2 days free float.

Where corrective action has to be taken the jobs where the time can be recovered can be seen very clearly. In all cases, the availability of resources has to be checked also.

Box 8. 'External' changes. In addition to changes of schedule caused by variations in performance, changes may also occur from external reasons. For example, a shipping date is advanced, or a competitor's action means that a launch date has to be brought forward. The replanning that is necessary under these conditions can be seen very clearly by examination of the schedule bar chart.

SETTING UP THE REPORTING SYSTEM

This needs to be done in as simple a way as possible. The men who have to execute the various jobs have enough on their hands without having to compile massive reports. In an actual case concerning the re-launch of food products, instructions were issued that progress had to be reported on a number of set dates and, as soon as it was known that delays were going to occur, that these delays should be reported also. The report form was similar to that shown in Table 7.1.

In another instance concerning the construction project, the site agent was provided with a planning board on which the jobs were set out on a number of cards cut to the appropriate length. On completion of a job the card was removed, the completion date inserted and the card sent to the control office.

IMPORTANCE OF CONTROL

Unless the schedule is constantly updated it soon becomes meaningless and will fall into disuse. If this happens, most of the benefit of the planning stages will be lost. At least as much attention must be given to the control and reporting procedure as was given to the planning.

Table 7.1. CONTROL AND REPORTING FORM—PROJECT x
(Situation at 15 March)

Job No.	*Description*	*Duration*	*Responsibility*	*Start Dates*			*Finish Dates*			*Float*	
				Scheduled	*Revised*	*Actual*	*Scheduled*	*Revised*	*Actual*	*Total*	*Free*
1	Prepare presentation	5	Marketing	8 Mar		8 Mar	15 [illegible]		15 Mar	0	0
2	Presentation	1	Marketing	15 Mar		15 Mar	16 Mar		16 Mar	0	0
3	Prepare brief and brief advertising agency	16	Marketing	8 Mar	15 Mar	15 Mar	1 Apr		8 Apr	19	0
4	Agency prepare plan	30	Marketing	1 Apr	8 Apr		13 May		20 May	19	0
5, etc.											
				These dates prepared from schedule bar chart	Agreed revisions at updating points		These dates prepared from schedule bar chart	Agreed revisions at updating points			

(By courtesy of Brown & Polson Limited)

UPDATING THE ANALYSIS BAR CHART

Most people will probably prefer to record progress on the schedule bar chart but the analysis bar chart can also be used for this purpose.

In order to provide a common starting point for analysis, a job box is inserted at the start of the network to indicate how much

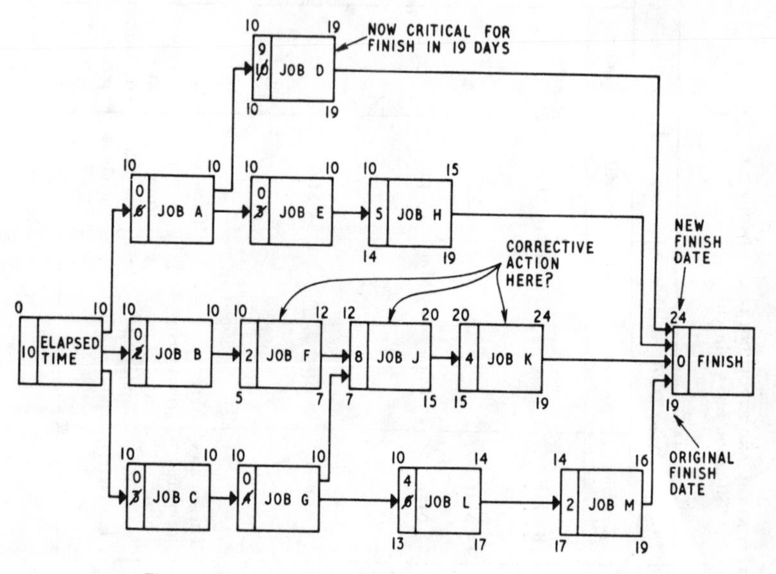

Fig. 7.3. Progress and plan compared on the analysis bar chart

time has elapsed. Jobs which have been completed are shown to have a duration of 0. Take the standard network as an example. Suppose that we are now at the end of day 10 and progress reports show the following situation:

Job	*Status*
A	Completed
B	Completed
C	Completed
D	Started—9 days' work remaining
E	Completed
F	Not started—2 days' work remaining
G	Completed
H	Not started—5 days' work remaining
J	Not started—8 days' work remaining

K Not started—4 days' work remaining
L Started—4 days' work remaining
M Not started—2 days' work remaining

(Note that for jobs that have not started the number of days work remaining indicates whether there has been a change in job duration.)

The analysis bar chart becomes as shown in Fig. 7.3. It can be seen that, when the analysis has been carried out, the original duration of 19 days has now been extended to 24. This is because Job F, which had a latest finish of 7 on the original network, will not now finish until day 12, hence the 5 days delay. The jobs where the time might be recovered are also easy to see, namely Jobs J and K. Unless something is done about their duration the project will be late. Note that Job D is also critical now, if we are to finish in 19 days.

SUMMARY

No one should get upset when networks and schedules have to be changed: this is what they are for. ABC is a useful means of comparing progress with schedule and helps management to decide on corrective action. Reporting systems need to contain the minimum amount of data consistent with adequate control. They need to be simple to operate. Updating can be carried out on the analysis bar chart or on the schedule bar chart.

8

The Use of ABC in Repetitive Manufacture

So far, the ABC method has been considered as applying to projects where every job in the network happens once only.

When there is a number of items to be produced, and each job happens more than once, ABC can be used in a different way. The method of application will be illustrated by an example.

THE HUNDRED UNITS

Suppose that a hundred units of an electro-mechanical equipment have to be delivered to a customer against the delivery schedule shown in Table 8.1. In this table, the delivery programme is shown extending over 30 weeks and the cumulative deliveries required are shown against the week numbers.

Suppose also that the equipment to be delivered consists of a

Table 8.1. ELECTRO-MECHANICAL UNITS—DELIVERY SCHEDULE

Week No.	*Cumulative Quantity Required*	*Week No.*	*Cumulative Quantity Required*	*Week No.*	*Cumulative Quantity Required*
0	0	11	31	22	65
1	2	12	33	23	70
2	5	13	34	24	80
3	7	14	36	25	85
4	10	15	38	26	90
5	12	16	42	27	92
6	15	17	44	28	95
7	19	18	48	29	97
8	20	19	50	30	100
9	22	20	55		
10	25	21	60		

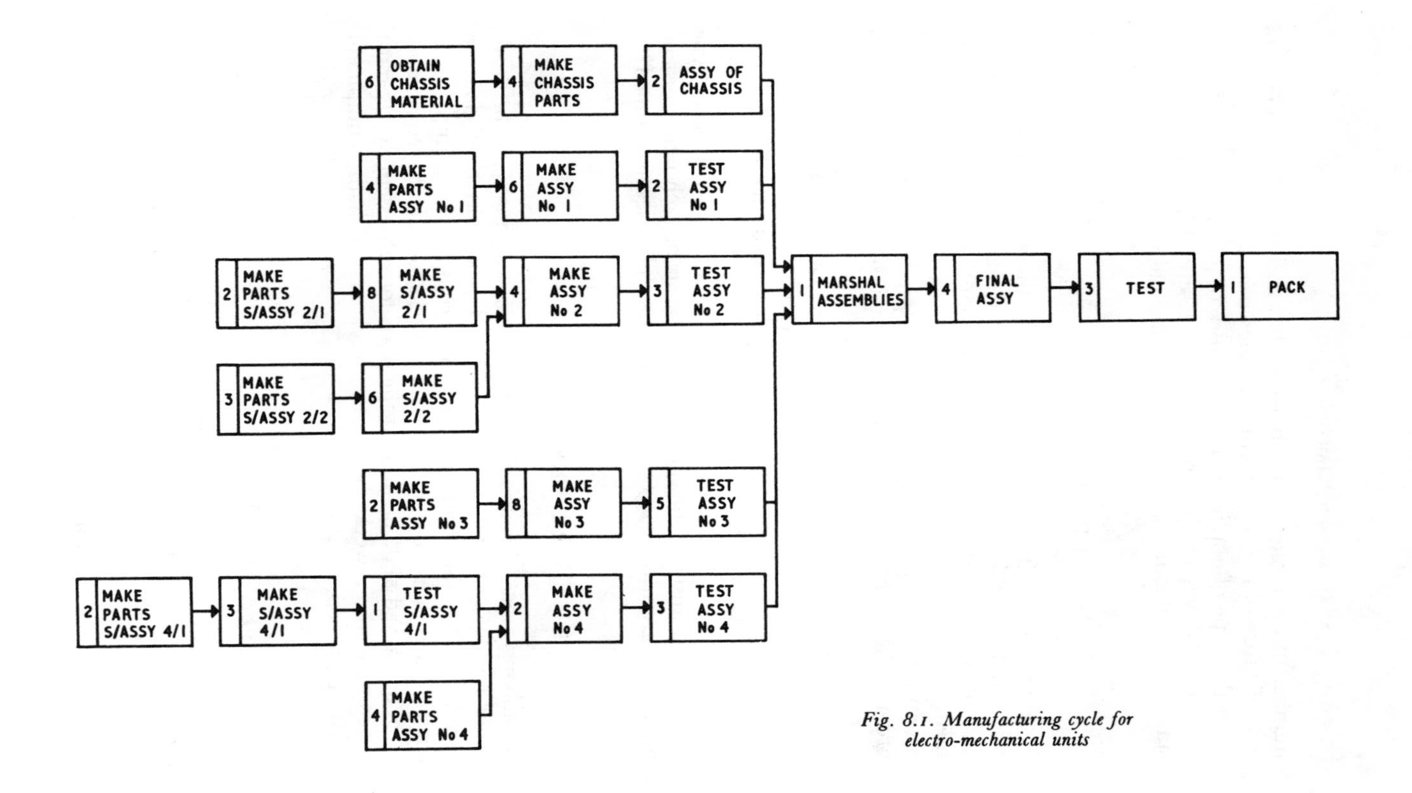

Fig. 8.1. Manufacturing cycle for electro-mechanical units

number of assemblies and sub-assemblies and that the sequence of manufacture of the equipment is as shown in Fig. 8.1.

In this type of application the network of jobs follows the 'family tree' breakdown of the product very closely. The times for each stage of manufacture are shown in the normal way.

THE PROBLEM

It is very clear from the schedule of delivery requirements how many units must be produced by a given week. For example, by the end of week 14, 36 must have been completed. What is not immediately clear is how many of the various assemblies, sub-assemblies and parts must have been produced by week 14 in order to keep the system 'balanced'. For instance, how many sets of chassis parts should have been made at week 14 in order to ensure that the future delivery programme can be maintained?

ABC, when used as a single-project planning tool, answers the question: '*By when* must each job be done in order to finish the whole project at a given time?'

Where it is used as a tool for planning repetitive work of the type described, it answers the question: '*How many* of each item must be made at a given time in order to maintain deliveries as planned?'

LEAD TIMES

By making a pass through the network from the final job, the 'lead' times for each stage can be determined. This has been done in Fig. 8.2. From this figure it can be determined, for example, that chassis parts must be completed 11 weeks before the equipment must be packed and, again, that the parts for sub-assembly 2/2 must be completed 22 weeks before the week in which the equipment must be packed.

STAGE TIMES

When ABC is used for repetitive manufacture in this way stage times need careful estimation. Anyone who has been concerned with production control will appreciate the difficulty of estimating these times.

When batch production methods are being employed, work

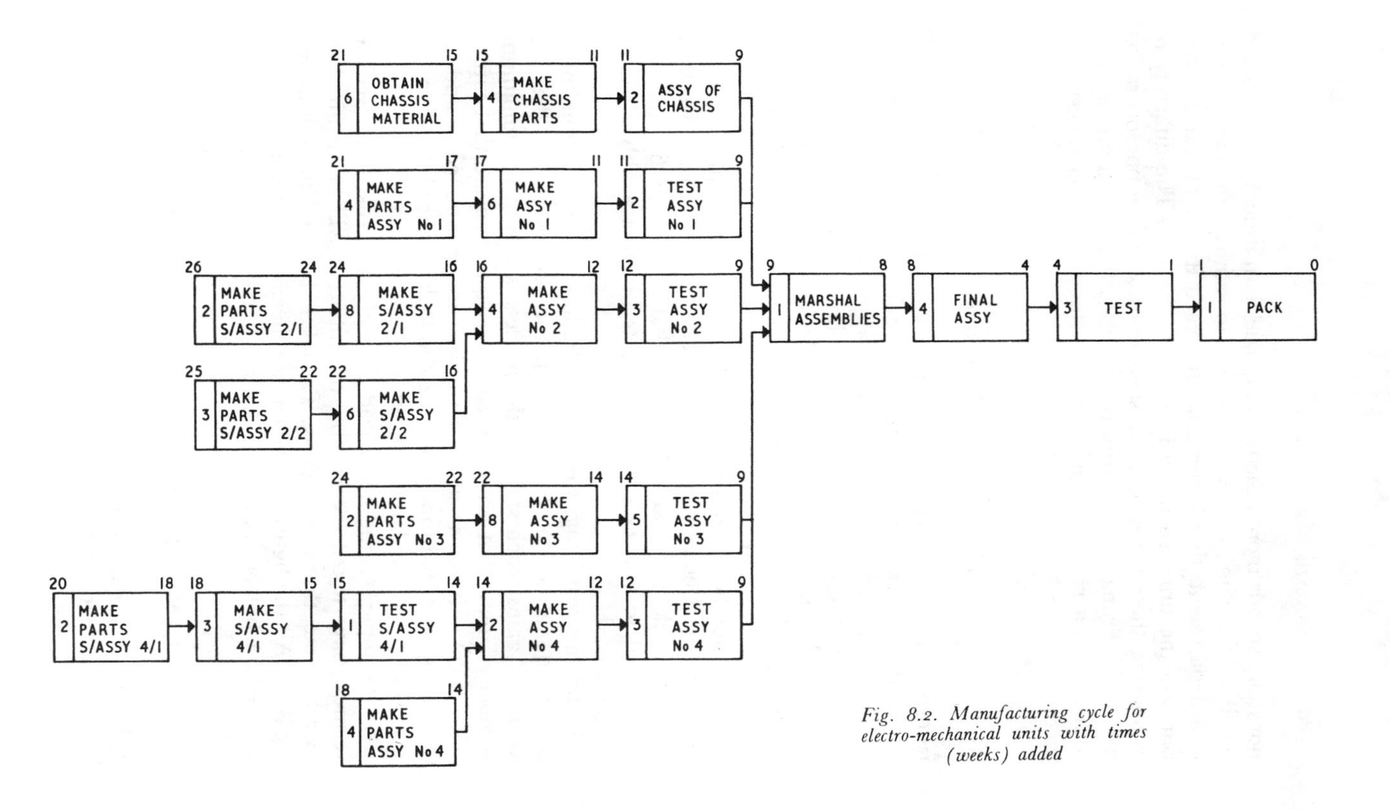

Fig. 8.2. Manufacturing cycle for electro-mechanical units with times (weeks) added

normally spends more time in the queue than it does being worked on. It is, therefore, necessary not only to estimate the process time in a department, that is the time that each item is operated upon, but also the time when the items are lying idle. The difficulty of estimating these times leads to such rules as 'allow one operation per week'. Probably the best way is to observe past achievements or to arrive at a realistic figure for each stage by direct observation methods.

MINIMUM BALANCE QUANTITIES

In order to find out the minimum number of items that must have passed through each stage by any given date, it is necessary to 'look ahead' in the delivery schedule by the appropriate lead time.

Suppose that we wish to determine how many sets of parts for sub-assembly 4/1 must be made by week 4. These parts must be completed 18 weeks before the unit they will be part of finishes the cycle, that is, before it is packed.

So the number of sets of parts is the number required for the units that must be delivered in 18 weeks' time, that is, at week 22 (18+4). Looking at Table 8.1 we see that at week 22, 65 units will be required, so at week 4, 65 sets of parts for sub-assembly 4/1 must be completed.

Using this reasoning, we can now build up a table of minimum balance quantities for each stage in the process (Table 8.2) for week 4. Against each stage in the process is placed the minimum number of items that are required at week 4 to ensure that future delivery commitments can be met. These are determined by adding the appropriate lead time to week 4 and looking ahead that number of weeks in the cumulative delivery table (Table 8.1).

For example, testing has a lead time of one week, so at week 4 we must have tested the number of units that will be required in 1 week's time, that is, at week 5. Table 8.1 shows us that this quantity is 12. All the figures in the column headed 'Minimum number required for balance', in Table 8.2, are found in this way.

'LINE OF BALANCE' CHARTS

If Table 8.2 is expressed graphically, the chart in Fig. 8.3 results. This is where the term 'line of balance' comes from. [LOB was certainly in use in the U.S. before the last war and was one of the ancestors of the present family of network techniques. Its use is increasing in the U.K. at present. There seems little point in separating it from network techniques since it is only an adaptation of the 'one-off' project approach.] Each of the stages must have produced at least the quantity shown on the chart if the delivery programme is to be met.

ALLOWANCES FOR FAILURES

Minimum balance quantities need to be increased if there is a risk of failure in the manufacturing process. Suppose that the manufacture

Table 8.2. MINIMUM BALANCE QUANTITIES—WEEK 4

Stage	*Lead Time*	*Minimum Number Required for Balance*
Pack	0	10
Test	1	12
Final assembly	4	20
Marshal assemblies	8	33
Assemble chassis	9	34
Make chassis parts	11	38
Obtain chassis raw material	15	50
Test assembly No. 1	9	34
Make assembly No. 1	11	38
Make parts for assembly No. 1	17	60
Test assembly No. 2	9	34
Make assembly No. 2	12	42
Make sub-assembly 2/1	16	55
Make parts for sub-assembly 2/1	24	95
Make sub-assembly 2/2	16	55
Make parts for sub-assembly 2/2	22	90
Test assembly No. 3	9	34
Make assembly No. 3	14	48
Make parts for assembly No. 3	22	90
Test assembly No. 4	9	34
Make assembly No. 4	12	42
Test sub-assembly 4/1	14	48
Make sub-assembly 4/1	15	50
Make parts for sub-assembly 4/1	18	65
Make parts for assembly No. 4	14	48

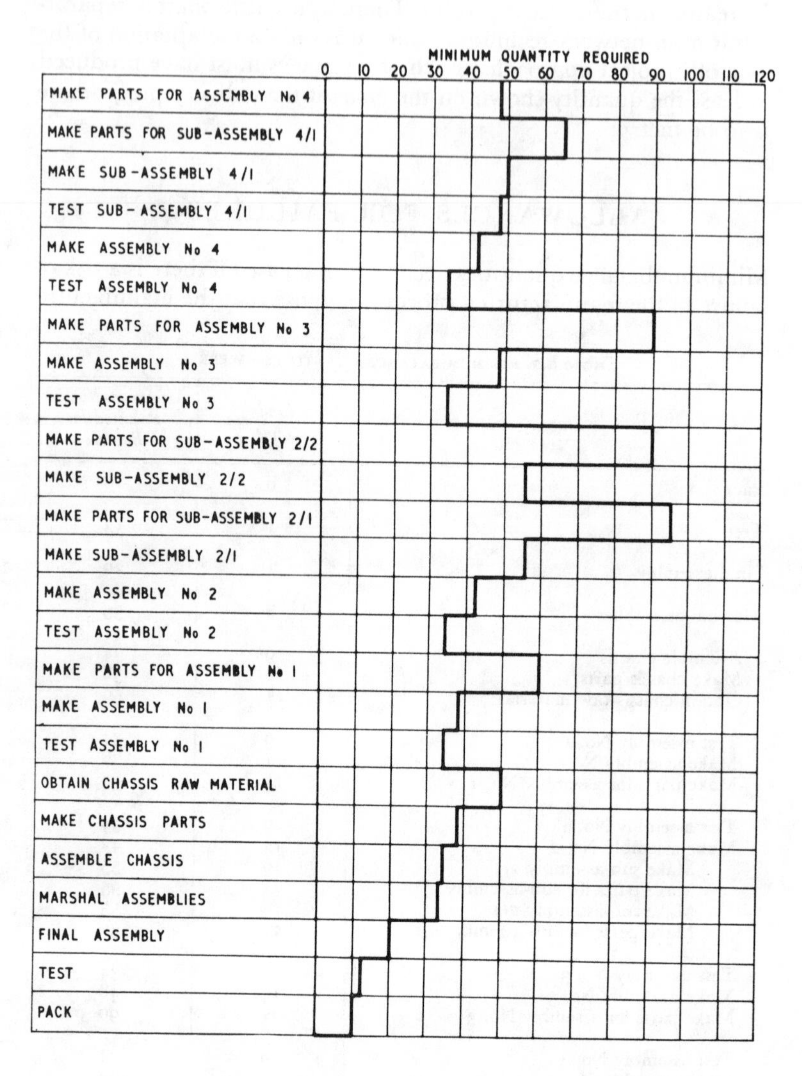

Fig. 8.3. The hundred units—line of balance for week No. 4

of assembly 2 is particularly difficult and it is estimated that 10 per cent of the assemblies will have to be scrapped. It will be necessary to make allowances for this by increasing the minimum balance quantities of assembly 2 by 10 per cent. The quantities of sub-assemblies 2/1 and 2/2 and the parts for these will also have to be increased by 10 per cent.

ALLOWANCES FOR RE-WORK

If faults in manufacture can be rectified by re-working, then minimum balance quantities do not need to be increased, but allowance for re-work should be made when the stage time is estimated.

CONTINGENCY ALLOWANCE

It is probably risky to work to minimum balance quantities in situations where uncertainty exists. Under these conditions a contingency allowance of, say, 10 per cent might be used to insure against unforeseen difficulties. The premium that has to be paid for this insurance policy is the inventory charges that arise as a result of the increased quantity of work in progress.

MULTIPLE COMPONENTS

If the same type of components are used in several of the assemblies or sub-assemblies of the equipment, it becomes necessary to aggregate the requirements. In the example, suppose that 3 of the parts used in sub-assembly 2/2 are also used in sub-assembly 4/1. By week 4 the minimum balance quantity of parts for sub-assembly 2/1 is 90 and for sub-assembly 4/1 is 65. The total requirement for the three common parts is thus $90+65 = 155$. When loading the machine that makes these common parts it is necessary to take this into consideration.

BATCH QUANTITIES

If on the grounds of economic batch quantity theory, or on some other grounds, it is necessary to manufacture parts in batches, the minimum balance quantities must be adjusted to take this fact into consideration. In this case, as soon as one of the parts is required the whole batch is required, and adjustment must be made accordingly.

For example, if it has been decided that the parts for assembly No. 1 are to be made in batches of 50, as soon as the first one is required 50 must be produced, and as soon as more than 50 are required, 100 must be produced. Initially, the stage time used should be that for one-off, but when it is decided to produce in batches, the stage time must be increased accordingly and the balance quantity recalculated.

PLANNING

The planning phase when ABC is used for repetitive manufacture consists of the following steps:

1. Draw the network showing the manufacturing cycle. This will normally follow the 'family tree' of the product.
2. Estimate the stage times and place them against the stages.
3. Make a pass from the last job through the network to work out the lead times.
4. Calculate the minimum balance quantities required in each week of the programme by looking ahead in the delivery schedule by the appropriate lead time.
5. Make allowances for scrap, re-work, multiple components and batching requirements.
6. Display the balance quantities for each stage in a table or as a line of balance chart. This needs to be done for each week of the programme.

CONTROL

When production starts and performance is measured, the line of balance chart serves as a useful means of comparing progress with plan. An example is shown in Fig. 8.4, where the dotted lines indicate performance and the solid lines the plan. It can be clearly seen how each stage is performing, and corrective action can be taken accordingly.

The variation from performance for each stage can be plotted weekly as a percentage of required performance, thus providing a

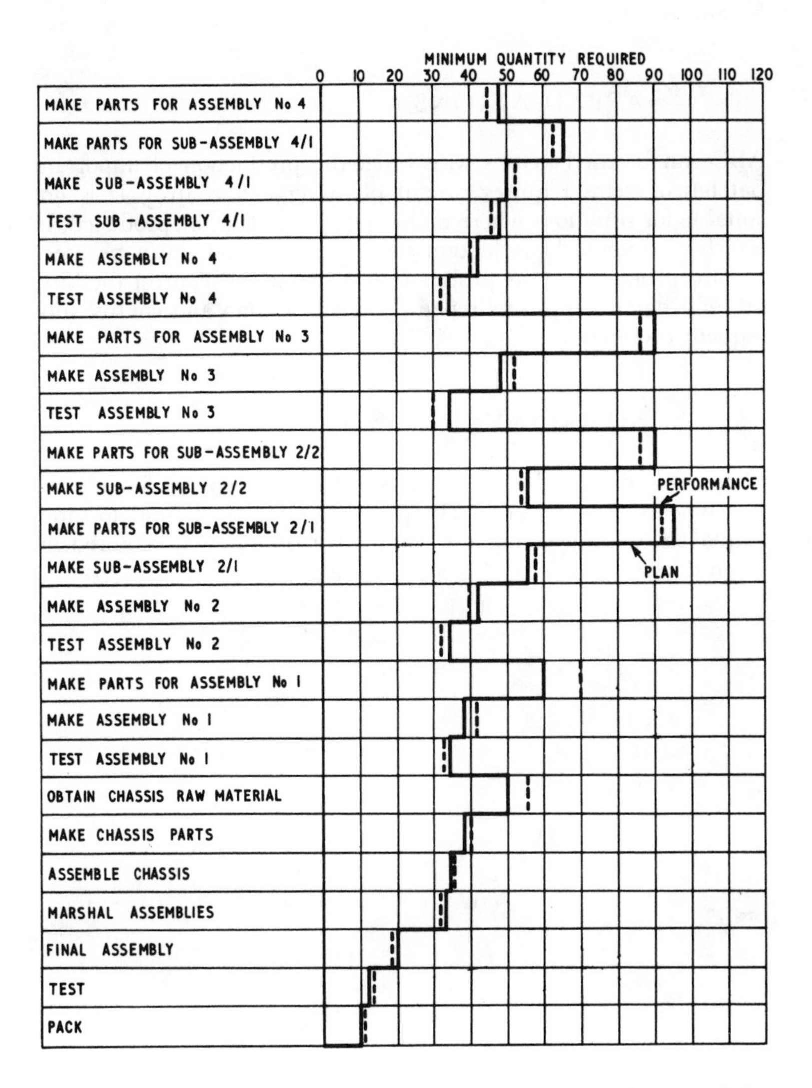

Fig. 8.4. The hundred units—line of balance for week No. 4 showing performance and plan

means of determining if the situation is getting better or worse and by how much.

APPLICATIONS

ABC can be used in this way when the production of important batches of items requires careful planning and control. It is not suitable for situations where many batches have to be produced, in which case normal production scheduling methods must be used. ABC can be used in its project form to plan and control the 'first off' of a batch and, in its repetitive form, to plan and control subsequent production.

SUMMARY

Planning and control of certain types of repetitive manufacture can be carried out using ABC. The planning steps are similar to those employed on projects but the emphasis is on quantity required by given dates:

9

Managing by ABC

The first thing that management has to decide about any network method, including ABC, are the reasons for using it. It is not sufficient to introduce a technique in the hope that *something* will be improved. The objectives should be thought through and, subsequently, management must determine if its objectives have been achieved. If not, the cause of failure must be determined.

OBJECTIVES

The usual reasons for adopting network techniques are:

- to provide a means of expressing complex plans and dealing with uncertainty in them
- to improve co-ordination and communication
- to determine priorities
- to reduce project times and improve time control
- to make better use of resources
- to provide better and more timely data for decision making
- to provide a means of ensuring that performance takes place in accordance with plans

SELECTION OF PROJECTS

ABC can be used on any project, but the selection of the first project that will be planned and controlled using it should be given careful consideration.

The importance of keeping networks simple has been stressed many times already and the same principle applies to the selection of the first project. Whenever possible, the first project should be a simple one. This will provide a gentle introduction for the managers who will be using the technique on more complex work later on.

It can be an expensive mistake to attempt a large-scale application of a network method without having built up confidence in it by earlier simple applications. An example of such a mistake occurred in a company who won an important contract. The contract was a very large one for them and involved the use of new technology. The management, wanting to ensure that there would be no failures to meet delivery dates, decided to use critical path analysis to plan and control the work involved in the contract. They had heard that CPA was the 'with it' thing for projects.

The combination of difficult technical work and a new and untried management technique proved too much for this company. The delivery date was not met and all confidence in network methods was lost, probably for ever. In addition to making the mistake of choosing a large and technically difficult project as a first application, this company had also made many of the other classical mistakes and became well and truly tangled up in its network.

It is possible, of course, to use outside experts for the first application of network methods and many companies have done this, thus buying other people's experience. But if this method is not used, it is important not to be too ambitious at the outset. Simple applications should be attempted first, followed by a more detailed planning and more precise control as confidence is built up.

AUTHORITY AND RESPONSIBILITY

A project needs a manager, someone who has the authority and responsibility to ensure that planning and control of the project takes place properly, who can take decisions and also make sure that they are acted upon.

Many applications of network methods have revealed that project management did not exist, because when the network had been constructed no one would admit responsibility for it!

Before planning starts it is essential to define who is responsible for the project and what his authority is. Most organisations consist of departments of one sort or another and it may be necessary to form a group of managers to plan and control projects through the system. One of these men, usually the one who is responsible for the largest part of the work, should be made responsible and can be given a title such as Project Co-ordinator.

Another method sometimes used is for the 'Chairman' of the groups to change as the project proceeds through the system. In the initial stages the representative from the design department is in the chair, handing over to the production representative when production starts; towards the end of production the marketing representative takes over, and so on.

It is usually a fatal error to appoint a junior member of staff to act as project co-ordinator. However competent he may be in the theory of network methods he is unlikely to obtain the co-operation that is necessary if networks are to succeed. He wanders around trying to get people interested, but no one pays any attention to him and his network is regarded as a useless gimmick. The network must be a manager's plan, not a planner's plan.

WHO DRAWS THE NETWORK?

The responsibility for producing the project plan rests firmly on the shoulders of the manager or group of managers who are responsible for the project. They must, therefore, be responsible for the logic of the network and for the estimates of the durations of the jobs. Since this is so, they may as well draw the initial network themselves.

One method commonly used is to assembly the managers concerned and to provide a blackboard or large sheets of paper. The network is then constructed, each man making his own contribution under the guidance of the chairman or project co-ordinator. This is usually a more effective method than employing one man to make a circular tour of the organisation to collect the information.

When the initial network has been drawn and time estimates obtained, it can be redrawn neatly and then analysed. This task can be carried out either by one of the managers or by a more junior member of staff appointed for this purpose.

It will then be necessary to hold another meeting to reduce the project time if this is required, and to make scheduling decisions so that the scheduled bar chart can be constructed. Again, the managers must make the decisions but the donkey work of drawing the scheduled bar chart can be delegated.

AMOUNT OF DETAIL

Where several levels of management are concerned it is advisable to construct networks corresponding to these levels. The senior management will be concerned only with major jobs and will require what are sometimes called 'summary networks' which show the work to be carried out in broad detail. At more junior levels of management more detail is required. One job on a summary network can be expanded into a network for use at these more junior levels.

By using this method, the amount of data is kept to reasonable proportions and managers are not confused by excessive detail with which they are not concerned. If excessive detail is presented the system will probably be abandoned.

A parallel can be drawn with the budgetary system in the organisation. The amount of detail in budgets varies with the level of management operating them. The same principles apply to project planning and control using networks.

THE COMMON MISTAKES

ABC is simple to understand but there are many snags and pitfalls to avoid in its application. These are not caused by the technique itself but are the usual management difficulties that arise when anything new is being attempted. The most common mistakes are as follows:

INADEQUATE SENIOR MANAGEMENT SUPPORT

Senior management needs to understand how network systems work, what they can do and what they cannot do. If they do not provide the drive necessary for successful application, failure is almost certain. Senior management must show that they have confidence in the system and must use it, believing the answers and facts that it produces however unpalatable these may be. For example, suppose that when a project has been planned, using ABC, its duration turns out to be far too long. Senior management's reaction must not be to say that the network is rubbish, but rather to instruct the people who have produced the plan to re-examine the network to find a way of reducing the time and the costs involved.

LACK OF TRAINING

The question of the type and amount of training required is examined in Chapter 10. Not much is required as ABC is so simple, but even this is sometimes neglected. One of the essentials is that everyone should co-operate and work willingly in the team. If some of the people involved do not understand what is going on they will probably be frightened by the system and may even actively oppose it.

Organisations who are getting good results from network methods have made quite sure that there is a widespread knowledge of the technique at all levels of management. This is considered to be so important by one particular organisation that formal training is undertaken down to operative level.

FAILURE TO USE THE SYSTEM

ABC will not succeed if it is treated only as an interesting appendage to the work. It must be the *only* means of planning and control and all levels of management must work through it and rely on it exclusively. All other methods of planning and control must be abandoned. To have two systems operating at the same time, ABC and conventional production control for example, is to invite disaster.

BAD PRESENTATION

Networks and bar charts must be neat and well set out. They must be readable at a glance and should not require extensive explanation by their authors.

FAILURE TO UPDATE

This point has been covered in Chapter 7, Control. At all times, the network and bar charts must display the actual situation. If they do not, management will lose confidence in them and start making progress investigations directly.

EXAMINATION OF RESULTS

When a network method has been applied it is necessary to conduct an examination to determine where it has had success and where it has failed. Have the objectives of applying it been reached? If not, then improvements to the system must be made. Where failures have occurred these will usually turn out to be failures in management rather than in technique. The system *will* work if it is properly applied.

COSTS OF APPLICATION

It is difficult, if not impossible, to answer the question of costs precisely. What does the present system of planning and control cost to apply? What is the cost of bad planning and control? Various estimates of the cost of network methods have been published, varying from a small part of 1 per cent of project costs, up to 2 per cent; but these figures mean very little. At the time of writing, the British Productivity Council is conducting an investigation into the costs of applying networks and their findings may throw some light on the matter.

It must be remembered that ABC will replace the existing planning and control system so it is the extra costs, if any, that are important. Usually, the gains in time and cost control far exceed the extra costs of using any form of network analysis. If network analysis is reduced to its simplest form, as with ABC, the costs are unlikely to be important.

SUMMARY

The most important aspects of managing by ABC have been dealt with. These are:

Objectives must be defined.
Simple projects are recommended to begin with.
Authority and responsibility must be defined.
A system of drawing the network must be evolved.
A hierarchy of networks may be required, each level having the amount of detail appropriate to the management using it.
Common mistakes have been explained.
Results of the application must be assessed.
Were the objectives achieved?
Costs of application are unlikely to be significant.

10

Training Requirements

There is little difference between the training requirements for ABC and those for conventional network methods, except that the former, being much simpler, need less time. For example, one does not need to teach the use of dummy activities.

MINIMUM CLASSROOM TRAINING

Since ABC is so simple, very little 'classroom' time is required. After a brief explanation of the four steps used in the planning phase, students should be pushed in at the deep end and given a series of exercises which should be designed to bring out all the relevant points. This is undoubtedly the quickest and most efficient method of getting the various points across.

Having produced schedules based on resource limitations built into the exercises, the project should be simulated by the issue of progress data so that students can practise updating methods.

To make life interesting, a number of spanners should be thrown into the works and students should be taken through the process of testing a range of courses of action to correct the situation. Exercises on project time reduction and costing should also be included.

Throughout this training it is important not to give the impression that ABC is a rigid technique that has to be applied as a 'drill'. It is an approach to planning and control using certain logical steps. It is up to the individual manager to devise any variation that he thinks might be appropriate in his case.

PRACTICAL TRAINING

The best way to learn about any technique is to use it in practice. After a brief period in a classroom the first project can be planned.

The difficulties encountered will nearly all be management ones; few technique difficulties will arise.

It is useful to have someone in the organisation who has had previous experience of applying network methods to projects. Provided he has learnt from his mistakes, they can then be avoided.

In Chapter 9, the main pitfalls were dealt with. In the absence of inside or outside experts, this will provide a good guide to practical application.

WHO SHOULD BE TRAINED

Anyone who is going to become involved with ABC in any degree needs to understand how it works. People who have to provide input data for the system would like to know how it is going to be used. If they understand this, their data will probably be more accurate.

Anyone who has to take action based on data output from the system needs to understand the source of this information. He will then have more confidence in its reliability.

The importance of senior management support was stressed in Chapter 9. At least one member of senior management should have an appreciation course, so that the capabilities and limitations of the technique are fully understood.

Training needs can be classified under three headings.

APPRECIATION TRAINING

One day in a classroom, with plenty of practical work, will suffice to give the required amount of knowledge about the basics of the technique. This course should be designed so that those attending will leave at the end of the day understanding what the people who are going to be using ABC are trying to do. When a course is designed for an organisation, the day should finish with an explanation of the planning and control procedures that will be adopted. In this way, those attending will not only understand the technique, but will have a good knowledge of the way it is going to be applied. This facility is not usually available on courses run outside the organisation and is a compelling argument for running courses 'in-plant'.

PRACTITIONER TRAINING

The people who are going to draw and analyse networks and produce schedules from them need another two days to practise the more detailed manipulations, such as resource allocation using histograms. Two hours should be spent discussing the planning and control systems and the difficulties likely to arise in practice.

The exercises used should include some based on the project work of the company, and if repetitive manufacture is to be planned and

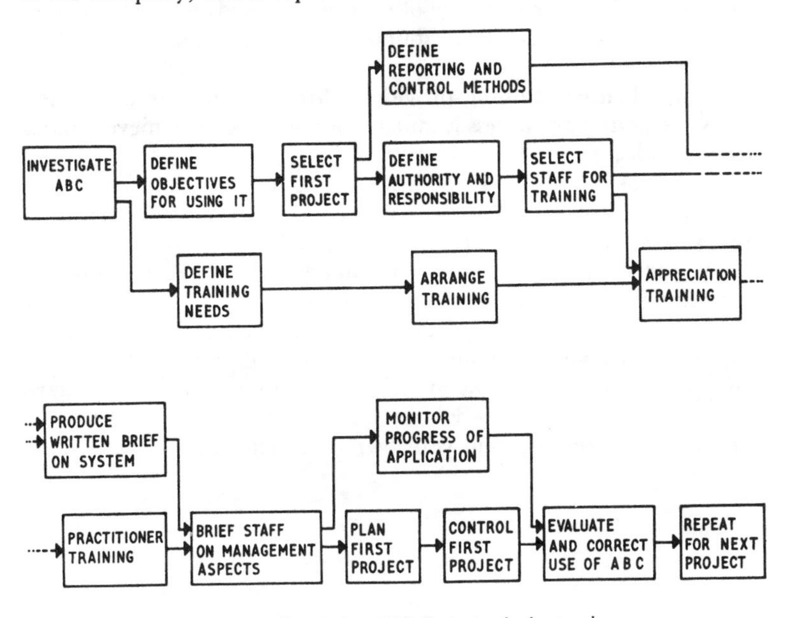

Fig. 10.1. Introducing ABC Basic standard network

controlled using ABC, then exercises on this aspect should also be included. Towards the end of the second day, work can start on 'real' networks using a team approach if this is appropriate. This provides a lead-in to the third stage of training.

TRAINING DURING APPLICATION

When work starts in earnest, it is useful to have someone available who can provide guidance in the technicalities and help to sort out the application difficulties that will arise. This is particularly important when the first updating point is reached. An expert opinion is also of value when the project has been completed and the success or otherwise of the system is being reviewed.

THE IMPORTANCE OF TRAINING

Organisations that have made the little training effort that is required to spread a knowledge of network methods throughout their management structure have found that this pays handsome dividends.

Comprehensive training provides a firm base for the use of network methods, and success is much more likely to be achieved under these conditions.

Successful training must be based on company policy for the use of networks and each company will have its own requirements. A generalised course is, therefore, of limited value unless the company representative who attends it can adapt what he has learned to the requirements of his own organisation.

Given a carefully devised training programme, a determination to get results and avoidance of the errors discussed in previous chapters, success is almost guaranteed. Figure 10.1 is a standard network for introducing ABC into an organisation. You can now go ahead and amend it to suit your own requirements.

Index